AF681334

BOXWOOD & IVORY

Stanley Traditional Rules, 1855-1975

Philip E. Stanley

Photographs by the Author

The Stanley Publishing Co./Westborough

Published by
The Stanley Publishing Company
P.O. Box 689
Westborough, Mass. 01581

Copyright © 1984 by Philip E. Stanley

All rights reserved. No part of this book may be reproduced or transmitted in any form or by any means, electrical or mechanical, including photocopying, recording, or by any information or retrieval system, without the written permission of Publisher, except where permitted by law.

Manufactured in the United States of America

Library of Congress Catalogue Card Number 84-050516
ISBN 0-9613291-0-6

First printing

Designed by Tomasino Capolongo

Dedicated To
CATHY, BARBARA, and GREG,
. . . *Who've had to put up with a lot while this was being written*

Acknowledgements

Another author/collector once told me, "What you're writing is the book you wish someone else had written before you started collecting Stanley rules". There's more than a little truth in that.

It started as a list. As I became interested in Stanley rules, I decided that I'd need a list to check them off as I acquired them. Gradually that list grew and evolved, first one line per rule, and then several, with the information becoming continuously more detailed. Eventually, about four years ago, it became an end in itself, and I decided to turn it into a book.

It would not be possible to have written this book without the aid and support of a great many people: tool collectors, librarians, historians, and other authors in the same or allied fields. They have given freely of their time and resources, often volunteering help even before it was requested; they have allowed me to photograph rules from their collections; and by publishing the results of their own research they have made available information and resources whose even existance was unknown to the author.

Particular thanks are due to four people: James and Ray Hill, Ken Roberts, and Roger K. Smith. Jim and Ray gave me access to the results of the years of research which they have done for Jim's forthcoming book on measuring instruments of all types. Ken, who has reprinted many Stanley catalogues over the years and authored numerous books and articles on tools, made available copies of a number of early Stanley catalogues. Roger not only gave me access to original Stanley catalogues in his collection, he also instructed me in the intricacies of book publishing. From the beginning all four have consistantly given me their support and encouragement, the most important factor in any project extending over several years.

A number of individuals have all allowed me to do research in their collections, and to photographs their rules. Without their aid the majority of the rules in this book would be listed with no accompanying illustration. They all have my eternal thanks.

William Baader	Springfield, Ohio
Aime Blouin	Georgetown, Mass.
Eugene Frankio	Brockton, Mass.
James Hill	Clarksville, Maryland
Charles & Walter Jacob	Pottstown, Penna.
Paul Kebabian	South Burlington, Vermont

Dave Knight	North Oxford, Mass.
Charles Leverone	Hanson, Mass.
Bud MacDonald	Wrentham, Mass.
Ruben Morrison	Brookline, Mass.
Kenneth D. Roberts	Fitzwilliam, New Hampshire
Roger K. Smith	Lancaster, Mass.
Bud Steere	North Kingstown, Rhode Island
John Treggiari	Dracut, Mass.

In addition to those mentioned above, the following are also deserving of special thanks: my designer, Tom Capolongo; my proofreader, Cecilia Thurlow; the librarians and staff of the public libraries of Worcester, Massachusetts, and New Britain, Connecticut; the directors and staff of the Connecticut Historical Society; the many members of the Early American Industries Association and Mid-West Tool Collectors Association who have helped me and shared their knowledge; and, last but not least, the nameless antique dealer who sold me my first rule (he has a lot to answer for).

To forestall questions, the author wishes to state that the name is purely a coincidence. He is not connected in any way with The Stanley Works, nor is he related to any branch of the Stanleys of New Britain.

The section THE ENGINEERS' (SLIDING) RULE, Chapter 5, in somewhat amended form, will appear in 1984 in the *Chronicle of the Early American Industries Association*

The section RULE ACCESSORIES, Chapter 4, in somewhat amended form, appeared in 1983 in *The Fine Tool Journal*

During its 130 year history the company whose products are the subject of this book has been known successively as A. Stanley & Co., The Stanley Rule & Level Co., The Stanley Rule & Level Plant of The Stanley Works, and The Stanley Tools Division of The Stanley Works. In this book the names Stanley, The Rule & Level, and The Company have frequently been used in place of these longer proper names to improve readability and keep the text as short as possible; no disrespect is intended or implied.

Excerpts from "History of the Rule Shop", from *The Rule Shop Seismograph,* I:5 (June 16, 1919): © 1919 by The Stanley Rule & Level Co. Used by permission.

Excerpt from "I Supervise the Making of Boxwood Rules", from *Tool Talks, No. 11:* © 1938 by The Stanley Works Inc. Used by permission.

Excerpt from *A History of The Stanley Works* by Robert K. Leavitt: © 1952 by The Stanley Works Inc. Used by permission.

Excerpt from *The Village Blacksmith* by Aldren Watson: © 1977 by Aldren Watson. Used by permission.

Table of Contents

6 Bibliography

CHAPTER 1 Stanley and the Rule Industry

Origins

The Stanley Tools Division of the Stanley Works has been manufacturing folding rules and measuring instruments for 130 years. Today, rules are only a small part of the Stanley product line, but in 1854, when the Stanleys first became toolmakers, rules were their initial (and only) product.

The numerous Stanleys in and around New Britain, Connecticut, had been involved with a large number of the manufacturing enterprise founded in that city during the period from 1810 on. Descended from a Thomas Stanley who had emigrated to New Britain from England during the early eighteenth century, they had participated, either financially or actively, in many of the mercantile concerns and hardware, clock, and lock manufacturing companies which had sprung up in that city since the end of the War of 1812.

In October 1854, four members of the Stanley family, Augustus, Gad, Henry, and Timothy, formed a partnership with Thomas Conklin, also of New Britain, for the purpose of manufacturing boxwood and ivory folding rules and other miscellaneous measuring instruments. Calling their firm A. Stanley & Co., after the senior partner, they commenced operations by purchasing the tools, machinery, and materials of Seymour & Churchill, a rule manufacturer in nearby Bristol, Connecticut.

Seymour & Churchill had been founded in the late 1840's by Henry Seymour, a former employee of Hermon Chapin, the rule and plane manufacturer of Pine Meadow, Connecticut, and Seymour's brother-in-law, John Churchill. The company was originally organized to manufacture clocks, but, this proving unsuccessful, Seymour fell back upon his earlier trade, and commenced the manufacture of folding rules.

It is not known whether Conklin, a former Seymour & Churchill employee, seeing the opportunity to purchase that company, went to the Stanleys for capital, or that the Stanleys, lacking rule manufacturing experience, sought out Conklin. Only three of the Stanleys were active in the management of A. Stanley & Co.: Augustus, his brother Timothy, and their distant cousin Gad. Henry, Augustus' and Timothy's older brother, while a substantial investor, was a silent partner.

Moving the Seymour & Churchill machinery into the upper story of the Sargent & Company factory in New Britain (owned by Henry Stanley), the new company immediately commenced operations. Manufacture of rules was begun, and one or more price lists, identifying the rules available by rule number, were issued.

Early Growth

In February 1855, after only a few months of operation, the company purchased the rule business of Seth Savage, a manufacturer in Middletown, Connecticut. The machinery thus acquired was moved to New Britain, and most of the Savage employees were retained on the Stanley payroll. This increased the work force to about eighty, and made it necessary to move operations to larger quarters, the upper story of North & Stanley's (another pair of relatives) hook and eye factory. Their 1855 price list (Ref. 1), issued after this acquisition, listed an increased line of 85 different boxwood and ivory rules.

Three years later the firm expanded again, and, more significantly, began to diversify their product line. Another New Britain concern, Hall & Knapp, had been organized in 1853 to purchase and operate the try square and bevel business of Hill & Crum of Unionville, Connecticut, and had since begun the manufacture of plumbs and levels as well. Henry Stanley was a silent partner in Hall & Knapp, as well as A. Stanley & Co., and in 1858, recognizing the complementary nature of their product lines, he arranged the merger of the two firms. The new company took the name of The Stanley Rule & Level Co., and the principals of A. Stanley & Co. were appointed as its officers, with Henry Stanley assuming the office of President. At the same time a new factory was built to house the rule manufacturing operations, across Elm Street from the former Hall & Knapp factory where the production of levels and squares continued uninterrupted.

This merger with Hall & Knapp was extremely valuable to Stanley in improving their business position and cash flow. A. Stanley & Co. had more than once been in very precarious financial condition, and had often had not enough cash in the till to pay the workers (Ref. 119). The resources and confidence necessary for the construction of a new 31 by 46 foot brick building were evidence of improved prosperity and sales, and the herald of the better times resulting from the merging of the two firms.

At this time the site of the new factory was far enough from the center of New Britain to be semirural. Behind the building to the west was a large pond called "The Canal," which provided water power to some of the local factories, and was sometimes fished in and skated upon by the Stanley workers. Across Elm street to the east were dwelling houses, with a large apple orchard (whose fruits were occasionally "sampled" by the employees) behind them, and a second, smaller pond beyond that.

Operations in the building were organized by floors:

In the basement the Try Squares and Bevels were made; the first floor contained the office, packing room and shipping room; on the second floor were the machine room, rule marking, blacking and polishing departments, and on the third floor the rule joints were made and all the press work. There was a small attic reached by a kind of step ladder stairs, where the paper boxes were stored.

There was an old fashioned elevator in which a large rope wound around a drum overhead, operated by anyone who needed it, but it worked so slowly that it was customary for any boy, who chanced to be on the third floor or attic of the building in which it was located, to slide down the rope, which saved time and shoe leather at the expense of clothing and accidents.

(Walter, Ref. 119)

In the spring of 1862, the company expanded once more, by purchasing a tool handle business owned and operated separately by Augustus Stanley in partnership with still another relation, John Stanley. This not only broadened the product line further, but also gave the company access to the machinery and skills needed to manufacture small turned-wood products such as chessmen, door stops, drawer knobs, etc.; in short, anything which would help offset the disruption caused in the tool industry by the Civil War.

E.A. Stearns & Co.

A year later the Stanley Rule and Level Company purchased a firm in Brattleboro, Vermont, E.A. Stearns & Co., whose reputation, manufacturing technology, and management were all welcome assets to Stanley's operations.

Sometime about 1833 or 1834, an S.M. Clark had started a rule-making business in Brattleboro, Vermont, which operated there for several years. By 1837 this business had shut down, and a year later one of Clark's former employees, Edward A. Stearns, purchased the factory, machinery, and remaining stock, and began the manufacture of rules under the name of E.A. Stearns & Co. Stearns was a stickler for quality, and his rules quickly acquired a reputation as the most accurate and best finished in the country. Much of their reputation for accuracy was due to the use of "special machinery" to graduate them, machinery invented by a Lemuel Hedge, of Windsor, Vermont, and perfected by him while a Stearns employee.

After the death of Stearns in July 1856, the Stearns company was partially or wholly acquired by Charles L. Mead, a native of Brattleboro, and continued in operation under his management. Mead was not a manufacturing man, however; his previous experience had been as a clerk in Massachusetts and as a bookkeeper in Rhode Island, and his primary interest was in sales and marketing. When the factory was partially burned out at or about the time of the Civil War it was not difficult for Henry Stanley to persuade him to sell out, and in 1863 he did so, accepting Stanley stock and the position of treasurer in the company in return for his interest. This acquisition was extremely beneficial to both parties; Stanley getting both the use of the Stearns machinery and the services of a first-class manager and salesman, while Mead, for his part, was able to concentrate on marketing, and leave the manufacturing end of the business to others.

At the time of the purchase Mead was then serving as a soldier in the Civil War, and it was judged best to leave the Stearns manufacturing operation, involving a work force of approximately thirty, in Brattleboro. In 1867, however, after his return, the machinery, together with many of the skilled workers, was gradually moved to New Britain, into a new building constructed for this purpose adjacent to the existing rule shop. This move was completed by about 1870. At the same time, Mead was placed in charge of the Stanley warehouse on Beekman Street in New York City, and concentrated on the sales end of the business.

The manufacture of Stearns rules was kept separate from that of the Stanley rules, and they were listed in their own section of the catalogue as "Stearns' Rules ... highly regarded for their uniform standard of accuracy, and finish." (Ref. 9). The extra pains

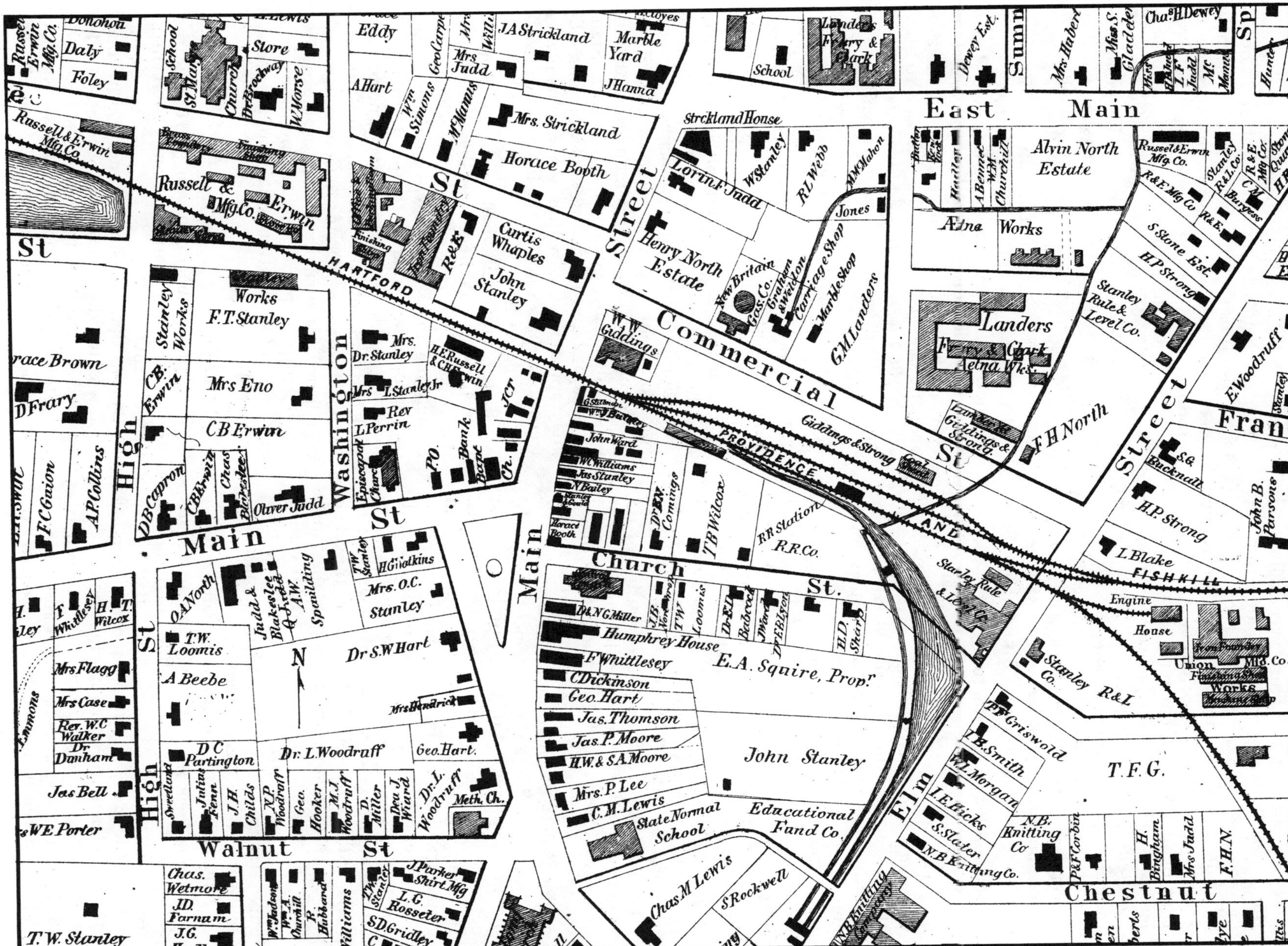

(Ref. 108)

Figure 1: New Britain and the Stanley Rule & Level Co. Factories in 1869

taken in their manufacture eventually made them prohibitively expensive, however, and between 1879 (Ref. 12) and 1884 (Ref. 13) all but four of the remaining forty-nine Stearns wood rules were discontinued, those four being retained, but renumbered as Stanley rules. The Stearns Ivory rules were manufactured until 1892 (Ref. 16), after which they also were no longer made.

Charles Mead had a long and distinguished career with the Stanley Rule & Level Co., succeeding to the presidency in 1884, and occupying that position until his death in 1899.

Bailey and Plane Manufacture

In the decade after the A. Stanley/Hall & Knapp merger the Stanley Rule & Level Company had been drifting, making a variety of disparate products in an effort to expand beyond the manufacture of rules, levels, and gauges. Many products were offered: stationers' supplies, checkers and chessmen, buttons, drawer knobs, toy pistols, sash pulleys ... even roller skates! For a brief period Stanley had acted as the selling agents for the products of other manufacturers, such as Brown & Sharpe's machinists' tools and Newark Eagle Tools' hatchets and axes. Many of these activities were profitable ventures, but they did not represent a focusing of the company's efforts on a single specific goal, and they tended to diffuse the efforts of the sales force by requiring salesmen to deal with a wide variety of establishments.

In 1869, Stanley management, aware of the problems which this lack of focus must have been causing, acted to remedy this. Deciding that from then on Stanley would concentrate solely on tool manufacture and nothing else, they took steps to eliminate all non-tool production, and to expand the product line with the addition of new types of tools. All extraneous products in the catalogue were eliminated at about this time (Refs. 8 and 9), and, in a step which was to have a great impact on the company, they purchased the small Boston plane manufacturing firm of Bailey, Chany & Co.

The planes and spokeshaves manufactured by Bailey, Chany & Co. were the inventions of Leonard Bailey, an erstwhile cabinetmaker from Winchester, Massachusetts. Bailey was one of the early inventors and designers of metallic planes, receiving his first patent (for a cutter-adjusting mechanism) in August 1855. Some time between 1855 and 1858 he had begun the manufacture in Winchester of planes and spokeshaves based upon this and other subsequently invented features. The design and finish of his planes was extremely good, but they must have been expensive to produce, since in 1862, and again in 1867, he redesigned them to reduce their cost.

In 1867–1868, Bailey, who had by then moved both his business and his residence into Boston, formed a partnership with an individual named Woods to manufacture planes under the name Bailey, Woods & Co., but this association lasted only a short time, and near the end of 1868 he formed a new partnership, Bailey, Chany & Co., with Jacob Chany of Medford, Massachusetts. It is likely that this pair of parnterships and the subsequent sale of Bailey, Chany & Col. to Stanley in 1869 were motivated by either management difficulties or a shortage of capital.

Stanley purchased Bailey, Chany & Co. in May 1869, moving the complete operation to New Britain in the same year, and supplementing it with new machinery. Leonard Bailey was hired to supervise operations, receiving 5 percent of the manufacturing cost of the planes and other tools he produced, and the following year "Bailey's Patent Iron and Wood Bench Planes, Veneer Scrapers and Iron Spoke Shaves" were offered in the catalogue (Ref. 8). The combination of Bailey's designs and quality control with Stanley's capital and marketing skill were immediately apparent, and within a few years planes were to surpass rules as the best-known items in the product line.

Bailey only remained with the Rule & Level for six years. Becoming dissatisfied with the royalty payments he was receiving, and contending that Stanley was marketing their own proprietary planes to the detriment of those in which he had an intrest, he terminated his contract with the company in June, 1875. Moving to Hartford, Connecticut, he established his own firm once again, at first manufacturing planes in competition with Stanley, and then, after 1882, copying presses and equipment.

Competition ...

The period beginning in 1870 was the zenith of rule manufacturing in central Connecticut.

By this date the small owner/journeyman shop was rapidly becoming obsolete, to be replaced by the large factory and mechanized production. Four firms, of whom Stanley was soon to become the largest, were beginning to dominate the rule market. Besides Stanley, the other three were:

H. Chapin's Son

H. Chapin's Son was founded in 1826 by Hermon Chapin, in partnership with Daniel Copeland, for the purpose of manufacturing woodworking planes. Their factory was in Pine Meadow, part of New Hartford, Connecticut. In 1828 Chapin bought Copeland out and became sole proprietor, selling his planes under the name of "H. CHAPIN – UNION FACTORY." In 1835 Chapin purchased the rule manufacturing business of Franklin Bolles of Hartford, Connecticut, and, moving the machinery to Pine Meadow, added rules to his line of planes.

Chapin operated his factory at Pine Meadow for 38 years, until 1866, a period during which rule manufacture steadily expanded, and rules became a more and more important part of his product line. A whole generation of apprentices were trained in rule making in his factory, and many of his craftsmen, such as Lorenzo Stephens and Henry Seymour, eventually left to form their own rule manufacturing firms.

Beginning in 1851 Hermon's sons Edward, George, and Philip joined their father in the business, and in 1860 it was renamed H. Chapin & Sons. In 1866, at the death of Chapin, Edward and George assumed ownership, operating as H. Chapin's Sons, but two years later in 1868 Edward purchased George's interest, and became sole proprieter, and the name of the firm was changed again, to H. Chapin's Son.

D.H. Stephens & Co.

Lorenzo Stephens worked as a rule maker, and his son Delos learned the trade of rule making while employed by Hermon Chapin at Pine Meadow. In 1854 they left Chapin and formed their own company, L.C. Stephens & Son, with Lorenzo as the senior partner. They must have operated in close cooperation with Chapin, however; they rented their first building from him, and for many years they sold him rules for remarking and sale under his own name (Roberts, Ref. 142).

Both Lorenzo and Delos were gifted inventors, interested in rules and the rule manufacturing process. In 1858 Lorenzo patented a combination rule combining the features of a folding rule with those of a level, bevel gauge, and square. Delos was more interested in manufacturing, inventing many machines for cutting the stock and making the plates and pins for the joints, and for graduating and marking the rules (Ref. 109).

In 1861, Lorenzo was succeeded by Delos as head of the firm, its name being changed at that time to D.H. Stephens & Co. Three years later, in 1864, the company moved all operations from Pine Meadow to Riverton, 8 miles away.

Belcher Brothers & Co.

The firm of Belcher Brothers was more than forty-five years old in 1870, having been started by a Thomas Belcher who was a rule maker in New York City as early as 1822. Together with his brother William he founded T. & W. Belcher in 1825, with Thomas in charge of rule manufacture and William handling sales. The name of the firm was changed twice during its lifetime, to Belcher Brothers in 1850, and then to Belcher Brothers & Co. in 1853. The offices of the company were in New York City, but their factory was located across the Hudson River in Camptown (near Newark), New Jersey.

Belcher Brothers manufactured a full line of folding boxwood and ivory rules, as well as board and log measures, gauging & wantage rods, tailors' squares, and other measuring devices, and was well known for its quality products.

At the time of the founding of A. Stanley & Co. Belcher Brothers was probably the largest single rule manufacturer in the United States, but the growth of Stanley and the other Connecticut firms gradually eroded this preeminence. By 1875 their market share had shrunk to less than ten percent, and in 1877 they had discontinued the manufacture of rules entirely.

The three largest of these companies, Stanley, Chapin, and Stephens, were all Connecticut-based, located within 30 miles of one another to the north and west of Hartford. Belcher Brothers, in New York/New Jersey, was experiencing declining sales and would cease operations within a decade. Some small shops, such as R.B. Haselton, of Contoocook, New Hampshire, and J. & G.H. Walker, of New York City, were still in operation, but their output was limited and their influence on the rule industry negligible.

Two other firms of significant size would eventually join the Stanley/Chapin/Stephens triumvirate of Connecticut rule makers, one located in nearby Farmington, the other in Cleveland, Ohio. These newcomers were:

Standard Rule Co.

Established in 1872 by Andrew Upson, president of the Upson Nut Co., the Standard Rule Co. was the youngest firm in this group of rule manufacturers. Standard's factory was located in Unionville (a section of Farmington), Connecticut.

It is not known if Standard began operations from scratch, or by purchasing the rule manufacturing operation of some other, smaller firm. Judging from the companys' rapid growth (within 3 years it had become the fourth largest rule manufacturer in the United States), the latter is more likely, but the name of the company bought out is unknown.

In contrast to the other Connecticut rule manufacturers, Standard used the same nomenclature and rule numbering scheme as Stanley. The other manufacturers, Chapin, Stephens, and Belcher, having begun manufacture before the Stanley began to dominate the rule industry, each had their own unique system.

E.T. Lufkin Board and Log Rule Mfg. Co.

The E.T. Lufkin Board and Log Rule Mfg. Co. was one of the large number of lumbering rule makers founded in the wake of the midwestern lumbering boom of the mid-1800's. Most of them quickly failed, due to the lack of necessary knowledge and skill, but Lufkin, founded in 1869, did not; using only the best hickory, and burning the figures into the surface for durability, this firm quickly achieved a reputation for quality.

The Lufkin factory was located in Cleveland, Ohio, and initially the company manufactured only log and board rules, in a wide variety of scales. This product line was subsequently expanded to include log calipers and other rules, but Lufkin remained primarily a maker of measuring devices for the lumber industry well into the 1890's.

The addition of Standard and Lufkin to this group of major rule manufacturers did not significantly affect the trend toward the concentration of rule making in central Connecticut. Lufkin, although doing well, was primarily a maker of board and log sticks for the lumbering industry, and was only competing with the Connecticut makers in a very small segment of the market; Standard, with its factory in Unionville, actually represented an acceleration of the trend.

By 1875 at least 95 percent of the rules made in the United States were being made within a 15 mile radius of Unionville, Connecticut (Roberts, Ref. 68). Of these Stanley produced 26 percent, Chapin and Stephens 22 percent each, and Standard 20 percent.

It is not surprising that such an industry as rule making should have become so geographically concentrated. The operations of Chapin, Stephens, Nelson & Hubbard, Solomon A. Jones, and other early Connecticut rule makers had created a work force in this area with the necessary specialized skills and a network of raw material suppliers that made it possible for new and expanding firms to prosper (a smilar example in our own time is the regional concentration, since the 1950's, of electronic-and computer-related industries near Boston, Massachusetts, and San Francisco, California).

... and More Growth

During the years 1854—1870, there were many changes to the Stanley line of rules. The merger with Hall & Knapp, and the purchases of the rule businesses of Seth Savage and E.A. Stearns had led to the addition and deletion of a number of rules in the successive catalogues. Of the 85 rules offered in the 1855 price list, eighteen had been discontinued by 1870, as well as two of the fifteen rules added to the list in 1859. Nine rules newly listed as "Stationers' Supplies" in the 1862 catalogue (Ref. 5) were discontinued within five years (Ref. 7), as were five other specialized rules for printers', hatters', tailors', and watchmakers' use. By 1870 Stanley still only offered 93 different rules.

In contrast to this early volatility, the thirty years beginning in 1870 were a period of slow steady evolution. The purchase of Bailey, Chaney & Co. had expanded the company's scope; Stanley was no longer a rule maker, but had become a tool manufacturer, and management's attention was divided between rules and other products. The rule market, relatively mature and stable, did not offer the opportunities for continued rapid growth that existed in the rapidly expanding metallic plane market, where the Bailey plane was just gaining acceptance. This tended to focus much of the management's interest on the development and sale of planes, and allowed the line of rules to stabilize and enter a time of orderly growth. Additionally, Stanley made no further acquisitions during this period; thus there was no necessity to integrate any other company's rule line with their own.

Of the 93 rules offered in 1870, 86 were still listed in 1898 (Ref. 17), together with sixteen additional rules introduced gradually, three or four at a time, over the intervening years. The only new products which did not succeed during this period were a pair of straight rulers introduced in 1874, and a group of eight metric rules which were offered from 1877 through 1879.

Stanley's sales of rules grew steadily during this period, both as a result of the gradual expansion of the market for rules, and at the expense of their competition. In 1875 the company was making nearly 26 percent of the rules in the United States. By 1877 this share had increased to almost 33 percent (Ref. 142). Exact figures are not available for subsequent years, but this pattern must have continued; in 1901 D.H. Stephens & Co. and H. Chapin's Son & Co., after Stanley the two largest rule makers in the United States, found it necessary to merge in order to continue operations in the rule market.

The Contract System

By about 1870 the rule manufacturing operation in the Rule & Level had been rationalized and brought under control by two trends: the development of the contract system, and the increasing use of automatic machinery.

The majority of the workers employed in the manufacture of rules, levels, and planes did not work directly for the Rule & Level, but for contractors who operated the various manufacturing departments. The contract system was an on-site subcontracting arrangement wherein Stanley furnished space, power, and materials, and the contractor provided the labor. Each contractor was responsible for a different phase of manufacture, or of a given type of tool, according to his area of expertise. Justus Traut, for instance, operated the department making combination and other special-purpose planes from 1875 through 1907 (Refs. 119 and 128); Thomas Conklin, one of the original founders of A. Stanley & Co. was a Stanley contractor (presumably in some phase of rule manufacture) for many years. The workers were employed by the contractor and paid by him (mostly on a piecework basis) out of the fee which he received from the company.

The contract system was apparently popular with the employees, particularly those who were highly skilled or trained. The contractor controlled not only wages, but working hours, and, to a certain extent, manufacturing methods as well. In many ways, working for a contractor was like working for a small company, retaining the close relationship between employer and worker which was often lost as an organization grew too large.

The system had both advantages and disadvantages for the Rule & Level. Neither Henry Stanley nor Charles L. Mead, who succeeded to the presidency in 1874, was a manufacturing man, and it was a great advantage to the company under their leadership to have a highly skilled body of contractors operating the manufacturing departments. On the other hand, the semi-autonomous status of the contractors made it difficult to effectively coordinate and schedule their activities or to establish uniform policies and working conditions (some contractors would allow their workers occasional time off for a baseball game; others saw nothing wrong with allowing an employee to send out for an occasion tin bucket [a "growler"] of beer).

Many of the contractors were Stanley's most prolific inventors. Traut, Henry Clark, James H. Eddy, and many others whose names appear on plane and tool patents during this period were contractors for various lengths of time. Being intimately familiar with the particular product of their department, and being skilled manufacturing men, they were constantly inventing new products and improving old ones, and developing new manufacturing methods to reduce their costs.

Automatic Machinery

The second trend, the increasing use of automatic machinery in rule manufacture, was particularly important in rationalizing production and keeping prices low in the face of rising labor and material costs.

The Rule & Level had managed to achieve this economy largely through the introduction of automatic machinery. The stock for rules was no longer shaped by hand planing, but was surfaced and finished to thickness and width by machines employing sanding drums. The brass plates and rolls for the joints were shaped using dies and other jigs, and the steel pins and rivets used in assembly were cut and driven by machine. The graduations, originally scribed on the rules by hand, one rule at a time, were now applied by machines which could process dozens of rules at a time. Only the figures, and some of the nonlinear scales, such as those on the gauging rods, were still put on by hand.

All of this did not happen at once, of course, but gradually, over a period of many years. In the words of John Sloan, a long-time (more than 50 years) Stanley employee who retired in the 1930's:

In most cases it was a hard job as there was very little improved machinery at that time. Most operations were done by hand. About 1880 I made the first jig for drilling hollows and about 1890 improved machinery began to appear.

(Ref. 124)

As these machines and jigs were introduced, and the manufacturing process was rationalized, the employees benefited as well. Most were on piecework, and even though the rates were reduced as productivity increased, they were still able to earn better than average wages. This was not universally true, however, and sometimes the workers would complain that the benefits of this automation were not being equitably distributed, with the company and the contractors getting the lion's share. Witness the following excerpt of doggerel written by one of the employees in February 1857:

... *When we the Rule trade first begun,*
We worked from morn, till setting sun,
And thought that when we had the trade,
We for it amply should be paid.

Perhaps there yet remaineth still,
If we will but abide your will,
Something that may us amply pay
For all the time we've thrown away.

If for the sake of gain, and trade,
A difference in your price-list made,
We all will with you rightly share
And thus help you the burden bear.

One reason why you do you say,
Is that we for improvement pay;
If that is so, we'll try the test,
And talk of them as we think best.

There's just one thing we want to ask
To help us bear the heavy task,
If after these old traps are paid,
Are we to have our wages raised?

Improvements have been made 'tis true,
To help the workmen as well as you,
But why must we for all these, work
To pay the maker while he shirks.

There is one machine been made of late,
That does add greatly to our fate,
Across the bench it lengthwise lies,
And looks immortal to our eyes.

It is a curious thing we know,
And down to Barnum's it should go,
And there among his trinkets, stand
The work of your machinist's hands.

When that tremendous thing was done,
To see it work we all did run,
But for our fun, we've dearly paid,
On it we've had our taxes laid ...

The Turn of the Century

In 1900, in spite of the growth of the Rule & Level during the previous thirty years, the organization of the company was basically unchanged. Henry Stanley and Charles Mead were both very much business executives in the 19th century mold, and the company reflected their attitudes.

... it was still the typical company of the 90's, when nearly all businesses were small and simple. Most of its correspondence was still conducted by hand, copies being made by the old letter-press device. For important letters there was one lady typewriting-machine operator—on half time. Few people used the one wall-type telephone instrument; it was considered safer (and sometimes was actually quicker) to go and talk face to face with anybody in town with whom you had business. The incoming mail was always opened in person by the Treasurer. No one else might touch it. Nor could anyone go through the stacks of newly-opened mail—no matter how important might be an eagerly-awaited letter—until the Treasurer had read it. Office hours were 8 to 12 and 1 to 6 every day, including Saturday. (The factory force worked 10 hours a day.) And in the whole office nobody was permitted to smoke except the manager, who rated the privilege of smoking fat Havana Seegars (sic) (with the bands on) not so much as a badge of dignity as to put visitors at their ease. ...

(Leavitt, Ref. 128)

The company's business was still primarily the manufacture of rules, planes, and levels. No other acquisitions had been made during the thirty years since 1870 (unless one includes the purchase of a few plane patents from their inventors to expand and round out the line of metallic planes).

Sales of rules and levels were satisfactory, but not spectacular. As Stanley had become the preeminent maker of these devices, dominating the market, there was less and less room to increase sales faster than the growth in the number of tool users. Plane sales were continuing to grow rapidly as the metallic plane found increasing acceptance among carpenters, and as more types of planes were added to the line.

The number of different Stanley rules had grown to 112 (Ref. 18), an increase of nineteen over the 93 offered in 1870. Thirty-eight new rules had been introduced (including four rules picked up under the Stanley name when the line of Stearns rules were discontinued after 1879), while nineteen had been discontinued.

By this time most of the small rule makers in the United States had either ceased operations or been bought out by the large Connecticut makers. A few one- and two- man shops, such as R.B. Haselton in Contoocook, New Hampshire, were still operating, but for all practical purposes virtually all rules made in the United States were produced by Stanley, H. Chapin's Son & Co., Stephens & Co., The Lufkin Rule Co. (by now located in Saginaw, Michigan), and The Upson Nut Co. (which had absorbed the Standard Rule Co. in 1889).

Growth Through Acquisition

When Frederick N. Stanley died in 1900 (only a year after his accession to the Stanley presidency upon Mead's death), he was succeeded in his turn by Charles E. Mitchell, a lawyer and a member of the Stanley board of directors. Mitchell, recognizing the need for company growth, and seeing new products as a means for achieving it, immediately embarked upon a program of diversification through acquisition. Under Mitchell's guidance, and that of Alix W. Stanley, who succeeded him after his death in 1916, the Rule & Level bought between eighteen and twenty firms during the two decades from 1900 through 1920 and for the first time became a broad-based manufacturer of woodworking hand tools.

The first such purchase was not of another company, but of the manufacturing rights for a new type of product, the Zig-Zag rule. This was an improved version of the so-called "folding lathe" offered by several English and American manufacturers, equipped with a spring joint to hold it rigid when open. The company's attention had been called to this new rule by Justus Traut, one of the Stanley contractors and a prolific inventor, and after buying the manufacturing rights from their German owners they added this new rule to the catalogue in 1900 (Ref. 18).

From then until 1920 other acquisitions followed in rapid succession. Bit braces were added to the line by the purchase of three small firms (Parmenter & Bullock, of Canada, and two others) in 1902, and a fourth, the John S. Fray Co., of Bridgeport, Connecticut, in 1905. Screwdrivers joined the line in 1904, with the purchase of Hurley & Wood, of Plantsville, Connecticut, manufacturers of the well-known "Hurwood" screwdrivers. In 1911 Stanley bought the vise business of P.J. Leavens, of Vinland, New Jersey, and began to offer his line of "Jersey" vises. Hammers and other impact tools came when the Rule & Level bought part of the defunct firm of Humason & Beckley, of New Britain, in 1912, and the Atha Tool Co., of Newark, New Jersey, in 1913. Framing squares were added in 1916, with the purchase of the Eagle Square Co., of South Shaftsbury, Vermont.

Not all acquisitions during this period were for the purpose of adding new types of tools to the line. Some, such as the purchase in 1907 of the Roxton Tool & Mill Co., of Canada, were to acquire manufacturing facililities in new market areas. Others were for the purpose of increasing manufacturing space in close proximity to sources of raw materials, such as the Harmon Woodworking plant in Ashfield, Massachusetts, bought in 1910. Still others were to buy out competing firms when they abandoned the plane business, such as the Gage Plane Co., of Vinland, New Jersey, in 1919, and the plane manufacturing operations of the Union Manufacturing Co., of New Britain, in 1920.

The sales of planes continued to grow at a healthy rate during the first two decades of the twentieth century. In the previous thirty years the metallic plane had become the standard tool of the woodworker, almost completely supplanting the wooden plane. Stanley, over that period, had expanded their line of planes to include more than a hundred different types, and had grown to be the pre-eminent maker of planes in the United States. This pattern was continued during 1900–1920, with planes being regularly added and improved, and by 1920 the company's production of planes of all types was more than a quarter of a million per year (Ref. 128).

Sale of traditional rules also continued to grow after 1900, but at a gradually slowing pace. The Zig-Zag rule had found almost immediate acceptance with carpenters, and this was soon reflected in reduced demand for boxwood rules. In spite of this declining demand, however, new rules continued to be introduced. Three 2 foot, 4 fold caliper rules, the Nos. 62C, 76C, and 83C, were added, as were a folding brass blacksmith's rule, the No. 17, and a whole series of folding and nonfolding patternmaker's shrinkage rules. By 1915 there were 132 listed in the catalogue (Ref. 31).

The other rule makers were feeling the effects of competition and the changing rule market more severely. The two largest, H. Chapin's Son & Co. and D.H. Stephens & Co., were both experiencing difficulties; by 1900 both had become almost exclusively rule makers, having by then lost much of the market for their wooden planes (their other primary product) to the Stanley-Bailey metal planes. By 1900 D.H. Stephens & Co. was approaching insolvency, a situation complicated by the fact that D.H. Stephens was getting old and had no close associates to share in the management of the company. Finally, in 1901, the two companies merged, Chapin buying the Stephens firm, and renaming the resultant combination, based in Pine Meadow, the Chapin-Stephens Co.

During 1900–1920 two other events occurred at Stanley which related to their rule making efforts: the introduction of a new wood to replace Turkey boxwood, and the end of the contract system.

By the turn of the century Turkey boxwood was becoming in shorter and shorter supply, and also more expensive. This was probably due to the exhaustion of readily accessible trees (Ref. 110), although it has also been suggested (Ref. 124) that this shortage was due to an embargo in about 1903. Whatever the cause, by about 1905 a substitute had to be found. The wood finally selected as most suitable was Maracaibo boxwood, a South American wood closely resembling Turkey Boxwood. The two species were only distantly related, but this new wood had nearly the same texture, color, and workability as the original, and, equally important, was available from Colombia and Venezuela in shipload lots.

The contract system, which had at one time flourished at the Rule & Level, had been less and less used after the turn of the century. As trained managers became available, one department after another was converted from contract operation to direct Stanley supervision, until by about 1905, only one remained under contract control. That one, the combination plane department, had been allowed to continue operating independently as a concession to the special status at Stanley of the contractor, Justus Traut. Finally, in 1908, Traut retired, and this last vestige of the early days at Stanley came to an end.

The years 1912–1915 represented the zenith of rule making at the Stanley Rule & Level Co. (see Figure 2). During the 60+ years prior to 1912 the line of rules had gradually expanded from 85 to 132 different types, including many specialty rules for professions and trades other than woodworking. This trend was not to continue, however. Already some of the rules requiring a great deal of hand work, such as the Nos. 6 and 16 Engineers' rules, had been dropped due to their cost. Other labor-intensive and low demand rules would follow, and then, as the Zig-Zag rule and the steel measuring tape became universally accepted, other rules as well. The dislocations caused by World War I, the Depression of 1929–1939, and World War II, would accelerate this trend, periodically causing wholesale cancellations, and by 1975 (Ref. 68) only four of these traditional rules would still be left in the Stanley product line.

World War I and Cancellations

The first of the three events to cause a major contraction of the product line was America's entry into the European war of 1914–1918. Even before the United States declaration of war in 1917, the production of goods for sale to the other belligerents was making materials such as boxwood and brass harder to obtain. After the event this problem became even more severe, and was further complicated by labor shortages caused by the military draft (99 employees, more than 8 percent of the work force, served in the military during the war [Ref. 128]). Additionally, Stanley accepted many contracts to supply standard and special tools, and other military products to the Armed Forces. The net result of all this was to necessitate major cutbacks in the production of many tools which were in limited demand or labor intensive, and the line of rules was one of the hardest hit.

Of the 132 different traditional rules and measuring devices offered by Stanley in 1915 (Ref. 31) nearly half had been discontinued by the end of the war. In 1917 alone, 29 rules

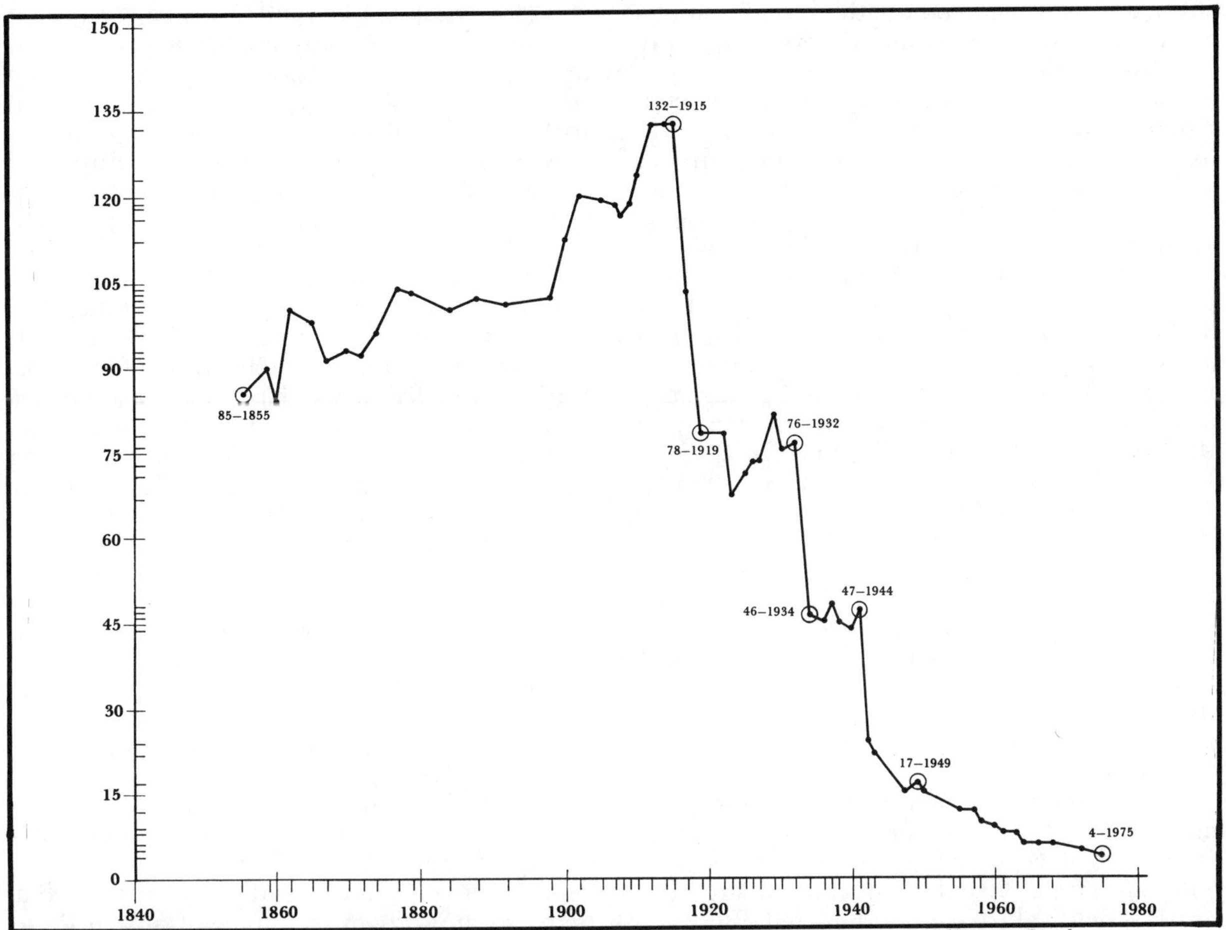

Figure 2: Stanley Rules; The Growth and Shrinkage of the Product Line *(Source: Stanley Catalogues, 1855-1975)*

were dropped (Ref. 32), including more than 20 which had been offered continuously since 1855. All of the board sticks were discontinued, and all but one of the rules with board tables. Five of the ten 2 foot, 2 fold rules were eliminated, including the three carpenters' sliding rules. One of the gauging rods (the No. 45½) was dropped, as were the Nos. 85, 89, and 92½ ivory rules.

Two years later this precipitous shrinkage of the rule line continued, with the deletion of 30 more from the catalogue. The twelve No. 31½ folding shrinkage rules were dropped, as were four more of ivory rules (the Nos. 86½, 88½, 90, and 91), and three more of the 2 foot, 2 fold rules. The No. 37 wantage rod was eliminated, as was the No. 66 4 fold yard measure. The one remaining rule with board measure tables, the No. 82, was discontinued.

Only 78 rules were still offered in the 1919 catalogue (Ref. 33), a skeleton of the broad product line of four years before. Even more significant, however, is the fact that none of the 59 rules dropped during the war was ever returned to production. The continuously increasing popularity of the Zig-Zag rule was cutting into the demand for boxwood rules to the point where the product line had to be pruned, and the rules dropped during the war were the resulting victims.

The Twenties

Probably the major event of the decade for the Rule & Level was its purchase in 1920 by The Stanley Works.

The Stanley Works, manufacturers of builders' hardware, was older than the Rule & Level (it had been founded in 1842, and incorporated ten years later), and more than twice as large (more than 3000 vs. 1252 employees) (Ref. 128). The two firms were closely related (as corporations go), with interlocking boards of directors, plants located close together in New Britain, and work forces drawn from the same labor pool (many local families had members working for both companies). Equally important, their product lines were both noncompetitive and complementary, with the same dealers frequently handling both.

The possibility of a merger had been considered for many years previously. As early as 1899, when there was some difficulty in selecting a successor to Mead, the directors of The Stanley Works had discussed the possibility of purchasing the smaller company. Finally, in 1920, an elaborate proposal was worked out involving the purchase of all Rule & Level stock in exchange for shares in The Stanley Works, plus cash. The operations of the two firms were not combined to any significant extent as a result of this merger; the Rule & Level simply became the Stanley Rule & Level Plant of The Stanley Works, and continued operations as before.

One result of this merger was the creation of the famous Stanley "Sweetheart" trademark. The directors of The Stanley Works decided that the company should have a new trademark to signalize the merger, and rather than simply design one arbitrarily, they elected to hold a contest, with the best suggestion for a new trademark winning a cash prize (Ref. 121). Guidelines were published, defining the features that the new mark should have, and the contest was opened to all Stanley employees. 320 contestants submitted 521 designs, and finally two nearly identical designs submitted by W.L. Hagen and E.C. Hartman were declared to be cowinners. The result was the famous "Sweetheart" trademark (Figure 3), combining the

(Ref. 34)

Figure 3: The "Sweetheart" Trademark, 1922

Rule & Level notched rectangle and the Stanley Works heart (adopted some years before to honor William H. Hart, for more than thirty years chief executive officer of The Stanley Works).

This trademark did not appear on all Stanley tools immediately. The preparation of the marking dies, labels, etc., took some time, and existing stocks of some tools had to be expended before new ones bearing the new mark would be made. It was not until 1922 that the order was given to begin marking Stanley rules with the new trademark (Ref. 141).

It was also in about 1918–1920 that Stanley ceased production of ivory rules. The cost of making these premium rules had been rising rapidly ever since the turn of the century, driven by higher labor costs (there was more hand work in an ivory rule than in an equivalent boxwood one) and an increasing scarcity of elephant ivory. By 1922, for example, the No. 87 2 foot, 4 fold ivory rule was priced at $78.00/dozen, a price guaranteed to discourage all but the most determined buyer! Nineteen twenty-two was the last year that ivory rules were offered; by 1925 the eight which had managed to survive World War I (the Nos. 38, 39, 40, 40½, 86, 87, 88, and 92) had all been removed from the catalogue, and Stanley was no longer a maker of ivory rules.

It is not known whether the ivory rules offered after the war were newly manufactured, or just existing stocks which were being sold out. It is significant to note that there are no known examples of any of these rules bearing the Sweetheart trademark. If they had still been in production in 1922, surely some would be found marked with the new trademark introduced in that year.

A few other rules were also removed from and added to the line during this decade. Two of the shrinkage rules were dropped, as well as the No. 44 wantage rod and the No. 42 ship carpenters' bevel, all in 1922. Fifteen new rules were added: one in 1923, four in 1925, two in 1926, and eight in 1929 (six of these last were very short-lived, however, and were removed from the catalogue only a year later, the first victims of the depression of 1929–1939). These were minor changes, however, and the 75 rules offered in 1930 were not a very different selection than the 78 in 1922.

More significant was the change in the character of the product line. Rules with printed figures and graduations (called "Heavy-Figure" rules) were introduced, the Nos. 61A, 66¼A, 66½A, and 68A (this last had actually been added to the catalogue in 1919). More rules with left-to-right graduations were cadded (the Nos. 73¼, 173E, and 173¼E), and others were switched over to this system of marking (the No. 71 and the three yardsticks). Meter sticks were offered (the Nos. 141 and 142), and for the second time in Stanley history all metal rules (the Nos. 450 and 550 counter measures). The traditional 2 foot boxwood rule with scribed/stamped markings was gradually being supplanted with a more diversified selection.

The 1920's also saw the departure from the traditional rule business of two competing manufacturers, the rule division of the Upson Nut Co. and the Chapin-Stephens Co., and the entry of a third, the Lufkin Rule Co. of Saginaw, Michigan. Both of the first two firms had been experiencing the same reduced demand for their traditional rules as Stanley, but unlike Stanley had neither the broad product line to absorb the slack, nor the resources of the Stanley Works to call upon. The Upson Nut Co., a Cleveland-based manufacturer of nuts and bolts, had retained its rule division (formerly the Standard Rule Co.) in Unionville, Connecticut, when Stanley had purchased its hand tool division in 1893 (Ref. 139), but finally, in 1922, abandoned this last vestige of toolmaking as well, and sold the rule division to Stanley. The Chapin-Stephens Company, whose primary products were boxwood rules and wooden planes, was similarly pressed by a shrinking demand for their output. A company reorganization in 1927 had not solved their financial difficulties, and in 1929 they finally ceased production. Stanley purchased the defunct company's line of rules, but only for the stock in process; except for Chapin-Stephens elaborate No. 036 combination rule, which was thenceforward marketed as the Stanley No. 036, no attempt was made to produce any of their unique rules, nor to continue marketing any ordinary rule under the Chapin-Stephens name.

Lufkin, on the other hand, with the revenues from its growing sales of measuring tapes, Zig-Zag rules, precision tools, and board and log sticks, began the manufacture of traditional folding rules sometime in the 1920's, in an effort to provide a broad product line for their customers.

The Depression and the 1930's

The economic depression of 1929–1939 was the second of the major events which were to accelerate the demise of the traditional rule at Stanley. An America where more than 30 percent of the work force was unemployed, and where industry was only operating at about 60 percent of capacity, could not continue to support a line of rules which was already under attack by newer types of measuring devices. Although Stanley probably suffered less than many firms (new construction might be at a standstill, but tools were still needed to repair the old), nevertheless many marginally profitable tools had to be dropped, and as at the time of the first World War, the traditional rules suffered worse than almost any other line.

It was not until after 1932 that the impact of the Depression was seen in the catalogue; like many other business leaders Stanley management did not at first believe that the economic slump was more than just a temporary phenomenon. In 1934, however, they faced reality, and eliminated thirty-two rules from the line, while at the same time adding two others, for a net drop of thirty. (There must be something mystical about numbers near thirty; in 1917 twenty-nine rules were dropped, in 1919 thirty more, and during the Second World War thirty more). A few other rules were added and deleted, two or three at a time, during the remainder of the decade, and by 1941 the 76 rules of 1932 had shrunk to 47. All four of the 1 foot, 4 fold rules were gone, as was one of the two 2 foot, 2 fold rules, and fourteen of the twenty-one 2 foot, 4 fold rules. The eight caliper rules had shrunk to five, and the eleven shrinkage rules to six.

In spite of the Depression, and notwithstanding this severe pruning of the line of rules, the 1930s continued to be a period of consolidation, growth, and acquisition for The Stanley Works. Four companies purchased in the late 1920's were consolidated into a new division, the Stanley Electric Tool Co. The Farrand Rule Co., of Berlin, New Hampshire, makers of steel "Pull-Push" rules, was purchased in 1930. Late in the decade the Company bought a controlling interest in the English toolmaking firm of J.A. Chapman, Ltd., of Sheffield. This was particularly significant in that it gave Stanley an entry into the English market, and, even more important, provided a plant where semi-finished tools, shipped from the United States, could be finished for subsequent sale.

As the company continued to grow and diversify, it was eventually decided to rename the division where the rules and other hand-tools were made. To refer to it as the "Rule & Level Plant" was to ignore the vast majority of the tools made by the various plants in Connecticut, Vermont, and England. As a result, in 1937 the name of what had been the "Rule & Level Plant of The Stanley Works" was changed to the "Stanley Tools Division of The Stanley Works."

It was also the 1930's which saw the end of the "Sweetheart" trademark. This occurred some time between 1932 (Ref. 43) and 1934 (Ref. 44); after twelve years some of the magic had rubbed off William H. Hart's name, and it was felt that a simpler trademark would be better suited for representing the increasingly broad line of products made by The Stanley Works. The resulting trademark, the plain notched rectangle enclosing the word "STANLEY" (see Figure 4), is the mark still used today. (This trademark is so often accompanied by the phrase "Made in USA" when marked on rules and other tools that it is frequently known by that name, in the same way that the 1921–1932 trademark is called the "Sweetheart" trademark.)

(Ref. 48)

STANLEY

Figure 4: The So-Called "Made in USA" Trademark, 1936

In 1934, for reasons which can only be surmised, Stanley made a significant change in the pattern of their caliper rules. Until 1932 (Ref. 43) it had been standard to make them left hand (see CALIPER RULES' Chapter 2), with right hand calipers only available as an option. Beginning in 1934 (Ref. 44) this was reversed; right hand calipers became the standard, and left hand calipers were the option. It is not clear why this change was made; the other major maker of boxwood caliper rules in the United States, the Lufkin Rule Co., continued to make its calipers left hand (Ref. 144) as late as 1940. Perhaps Stanley detected a trend in demand for the two types of patterns, with orders for the optional right hand caliper continually increasing; perhaps it was intended to make the Stanley caliper rules more acceptable in the English market, where right hand calipers were the norm (Refs. 16 and 19). There is no way of knowing.

World War II and the End of Traditional Rule Manufacture

The third major contraction of the line of traditional rules took place during World War II. Like all of American industry, The Stanley Works was called upon to divert almost all of its resources and production to war material. In the case of the Stanley tool divisions this was less of a dislocation than for many other companies; the same tools which they produced in peacetime were required by the war effort. The company was called upon to produce other products (rifle clips, etc.) as well, however, and this with a reduced workforce (1285 Stanley employees served in the armed forces between 1941 and 1945).

The net result of all this was to force the elimination, either permanently, or for the duration, of a great number of products, including another thirty rules. This was done gradually, in a series of steps (Refs. 52, 53, and 54). In 1942 eighteen rules were announced as either temporarily or permanently suspended, and in 1943, five more, leaving only twenty-four still in production. Ten of these cancelled rules were to return after the end of the war, in 1947 (Ref. 54) and 1949 (Ref. 55), but seventeen more were dropped at the same time, and less than a third of the prewar product line was still in existence when the war was over.

The seventeen rules that remained were the most prosaic in the line, the basic, high-volume rules which could (hopefully) withstand the competition of the Zig-Zag and Pull-Push rules which by this time were the carpenter's standard. Six were 4 fold rules, the Nos. 7, 27, 61, 63, 66½, and 68. Five, the Nos. 36, 36½L, 36½R, 136, and 136½, were caliper rules. One 2 fold rule remained, the No. 18, and one yardstick, the No. 41. The four remaining rules were two bench rules, the Nos. 34V and 34¼V, and two measuring sticks, the 4 foot No. 71 and the No. 510. In thirty-four years the line of traditional rules had shrunk to an eighth of its former size. Even more sad, all of the premier rules were gone: the ivory rules, engineers' rules, and board sticks; the wantage rods, Gunter's scales, and hatters' rules; the grandeur and diversity of the line was only a shadow of its former self.

Nor were most of those remaining seventeen rules to be in production for long. By 1975 only four were left. The other thirteen had, one by one, also been discontinued, either before or after the transfer of all rule manufacture from New Britain to the new Eagle Square plant in South Shaftsbury, Vermont. The last 4 fold rules, the Nos. 61 and 68, went in 1960, the last extension stick, the No. 510 (by then a Lufkin extension stick marked with the Stanley trademark), in 1972, and the last yardstick, the No. 41 (in continuous production since 1855), in 1975. After 1975 only four rules were left: two small caliper rules, the Nos. 136R and 136½, and two bench rules, the Nos. 34VR and 34½ VR.

CHAPTER 2 Materials, Construction, and the Rulemaking Process

The folding rules made by the Stanley Rule & Level Co. during the second half of the nineteenth century are wonderful examples of traditional form tools as manufactured by the evolving factories of that era. They used the same selected types of materials as when made by hand, and were constructed to the traditional patterns, but were made in lots of from 50 to 100 rules at a time. The level of mechanization employed primarily replaced human power with machine power, but did not significantly alter the fundamental sequence or nature of the operations performed.

Materials

These rules were made of three types of materials: wood, ivory, and metal. Wood was the most common, with ivory reserved for premium-quality rules, and metal only where its special properties were essential. The trim on the rules (the hinges, tips, etc.) was always metal, usually brass, but sometimes German silver, and occasionally steel. Hinge pins and the small pins joining the various parts together were either brass or steel.

Types of Wood

A number of different woods were used for Stanley rules. The most common was boxwood, but sometimes, for reasons of cost or physical characteristics, maple or hickory was used instead. Occasionally fancy woods, such as rosewood, satinwood, or mahogany, were used, but this was rare.

The preferred woods were light in color, and very fine grained. Coarse grained woods, such as hickory, were harder to graduate, as the lines and figures had to be cut deeper in order to be adequately visible. Dark woods such as mahogany and rosewood presented even more severe problems. An incised line filled with black paint or wax (the traditional method for marking wood rules) is almost invisible on a dark surface. Using a white filler is not a great improvement either; the contrast is not nearly as great and becomes even less as the rule becomes dirty with use and wear.

(Turkey) Boxwood

(Turkey) Boxwood *(Buxus balearica)* is the traditional, premium material for rule manufacture. It is a hard, heavy wood, with a creamy yellow color, and is among the finest textured of commercial woods, being almost grainless. It is an excellent wood for turning (such as chess pieces) and detailed carving (such as wood engraving). When properly seasoned, boxwood is dimensionally stable and insensitive to ambient humidity, and it is this property, coupled with its ability to be clearly and visibly marked with fine graduations which make it so desirable as a rule material.

Turkey box is a small tree (up to 36 feet high) which occurs locally from southern Europe through Turkey to Iran, the majority growing in Turkey and southern Russia. The logs of the tree are small, with an average diameter of from six to seven inches, and will sometimes have an irregular grain pattern.

Until the turn of the century the primary source of boxwood was Turkey, which was the shipping point for not only its own timber, but also that originating in Russia. The timber had become increasingly scarce and expensive, however, due to exhaustion of mature trees in areas accessible to transportation. It thus became necessary for American rule makers to find a substitute, and Maracaibo boxwood, or zapatera, a Venezuelan timber which had recently become available, proved to be nearly as satisfactory and was adopted.

The use of the European variety was never resumed. Improved transportation made new supplies available, but political and economic turmoil in the countries of origin had raised its cost to the point where the supply from South America was not only cheaper, but also more reliable.

Maracaibo Boxwood (Zapatera)

Maracaibo Boxwood *(Gonypiospermum praecox),* also known as zapatera, closely resembles boxwood, having the same fine, even texture and an only slightly lighter color. It has a straight grain, and, like boxwood, almost no figure. It is about 20 percent lighter than boxwood, and not quite so hard, and is thus marginally easier to work (although it splits more easily).

The zapatera is a small tree, similar in size to the Turkey box, which grows in the southern Caribbean. It grows in Cuba and in the Dominican Republic, but in large, commercially useful quantities only in Colombia and Venezuela. This wood has an advantage over boxwood in that it is available in slightly larger sizes (logs from this tree have a typical diameter of from 6 to 10 inches), and is less likely to have irregular grain.

Hard (Rock) Maple

Hard Maple *(Acer saccharum),* sometimes

known as Rock Maple, is a hard, dense wood, very pale, and usually straight-grained (although examples with "birds-eye" or "fiddle-back" figure are not uncommon). When straight-grained it will accept a smooth finish, and when thoroughly dry is reasonably stable. Hard maple has exceptional wear resistance (hence its use for bowling alleys, shoe lasts, etc.). The wood has a fine, even texture, and will easily accept fine graduations.

Hard maple is widely distributed in the North American continent, being common throughout Canada and the northern and central parts of the United States. The tree is medium to large in size, and its wood is available commercially in good-sized logs. Due to its ready availability, hard maple is a relatively inexpensive wood.

Hickory

Hickory *(Carya spp.)* is a tough, heavy, coarse-grained wood produced by four species of North American trees. The sapwood is almost white, the heartwood a reddish brown. The paramount quality of hickory is its high strength and impact resistance; it is frequently used for the handles of impact tools, etc. The coarse, ring-porous texture of the wood makes it difficult to mark fine graduations thereon, but it was still sometimes used for rules and sticks likely to be subjected to rough usage. Hickory is very dense, and seasons only very slowly, with much shrinkage.

Hickory is a medium to large size tree, indigenous to southern Canada and the eastern and southern United States. If properly grown, the logs are of good size, and the grain straight.

Mahogany

Mahogany *(Swietenia mahogani* and *swietenia macrophylla)* is one of the world's great cabinet woods, extensively used for fine furniture ever since the middle of the eighteenth century. Medium to dark red-brown in color, it exhibits a fine, even grain, and occasionally a spectacular figure. The Cuban mahogany *(S. mahogani)*, native to the Caribbean islands, was a darker and heavier wood than the Honduras or American mahogany *(S. macrophylla)* found on the Central and South American mainland.

Mahogany is easily dried and when properly seasoned is dimensionally stable and resistant to decay and insects. This, coupled with its ability to be easily worked to a high finish and to be carved with fine detail, has made mahogany probably the most widely used premium wood. In addition to fine furniture manufacture, mahogany is used in boat-building, pattern-making, and wood-engraving.

The mahogany tree is quite large, often attaining 150 feet in height, approximately half of which is trunk bare of branches. This trunk can be as large as 6 to 8 feet in diameter, so large as to require reduction to timber at the site before shipment to the ocean for transport overseas.

It was probably the Honduras mahogan that was used for rulers by Stanley in 1862-1865; by that date the Cuban variety was very scarce, and would have been much too expensive for use in commercial articles.

Rosewood

Rosewood *(Dalbergia spp.)* is a tropical wood of spectacular appearance, frequently used for cabinet work, as well as other purposes. A number of varieties exist (Indian, Brazilian, Honduras, etc.), but all have similar characteristics. The heartwood is purplish brown, with dark, almost black markings creating an exceptional figure. The sapwood is much paler, and is frequently discarded. The wood is heavy, and possessed of natural oils which enable it to slide easily on smooth surfaces (hence its occasional use for plane stocks and soles). Rosewood turns and carves fairly easily, and is frequently used for knife and tool handles (until the 1950's Stanley used rosewood for the handles and knobs on their planes, and for the stocks of squares, bevel gauges, etc.). Rosewood has the disadvantage of being difficult to glue, due to its oiliness, but in spite of this is frequently used for veneering because of its appearance.

The rosewoods are small- to medium-sized trees, widely distributed in the tropical regions of the world. The various species are found in commercial quantities in India, Madagascar, and Brazil, as well as Honduras and other Central American countries.

(West Indian) Satinwood (Yellowwood)

West Indian Satinwood *(Zanthoxylum flavum)* is a fine cabinet wood, one of a large number of different woods of African and Caribbean origin which have been loosely grouped under the name Yellowwood. Found in Puerto Rico, Santo Domingo, Jamaica, and in some of the smaller Caribbean islands, it had been a popular furniture wood in England 100 years prior to the popularity of

mahogany in Georgian times, and was still occasionally used as late as the end of the nineteenth century.

The wood is pale yellow to cream in color when cut, subsequently turning slightly darker with age. It has a fine texture, and is found with both straight and irregular grain, the latter producing an exceptional figure. Satinwood is a dense and somewhat oily wood, and gives off an odor similar to that of coconut oil when freshly worked.

The species can develop as either a shrub or a tree, depending on local conditions; when the latter, heights of 40 feet are common, and logs can be up to 20–24 inches in diameter.

Ivory

Ivory is the dentine-based living substance which forms the tusks of elephants and certain other mammals. It is a slightly creamy, almost white material, very dense, with fine close pores containing a gelatinous solution which renders it easy to work and polish. It is tougher than bone, but not so tough as horn, and will break under sufficient strain or shock.

Almost all ivory comes from elephants, the other mammals having much smaller tusks, and most of that from the African elephant, whose tusks are much larger than the Indian and Ceylonese species. Some ivory also originates as fossil ivory, being the tusks of mammoths found frozen in northern Siberia. At the peak of the ivory trade (1880–1890) 750 tons of African ivory a year were entering the world markets, about 60 percent through London, and the remainder through Bombay, Amsterdam, New York, and other cities (Ref. 110).

The properties of ivory which rendered it valuable (aside from its scarcity) were its hardness, its creamy whiteness, and its ability to be finely carved and polished. Until the invention of celluloid and other plastics, it was much used for billiard balls, cutlery handles, the faces of piano keys, and decorative pieces. Its hardness and brittleness required special techniques to work it, and it would gradually yellow with age, especially if kept away from the light. Ivory shrinks to some extent while drying, and must be seasoned like wood (it is probable that the "bleaching" process described by Stephens [Ref. 145, p. 376] was actually to ensure that the ivory was fully seasoned before making it into rules; sunlight will not bleach ivory, only slow down the rate at which it yellows). Even after seasoning it is not as stable with varying humidity as boxwood.

Types of Metal

Three types of metal were employed in the manufacture of the Stanley rules. Brass was the most common, but German (nickel) silver was usually used in its place on ivory rules and steel, of course, was used where high strength or wear resistance was required (for joint pivot pins and assembly pins, for instance).

Brass

Brass is an orange to yellow alloy of copper and zinc (sometimes with the addition of small amounts of other metals) which is almost ideal for the manufacture of the joint and trim parts of folding rules, and had been used for that purpose long before the Stanleys entered the rule business in 1854.

Brass is not as cheap as steel, but is still relatively inexpensive, and it has a variety of other properties which are highly desirable. It has a fairly low melting point, and can readily be cast. It is ductile enough to be easily rolled and drawn, and it machines freely. Brass will accept a high polish, which can be preserved by subsequent lacquering, and when so polished presents an extremely attractive appearance.

The wear resistance of brass is not exceptional (hence the number of well-used folding rules with very loose joints which are encountered), but its corrosion resistance is fairly good, so that, while it will tarnish, it does not have the extreme tendency to rust encountered in steel.

The brass alloy used for the joints and trim of the Stanley wood and ivory rules was a so-called "high" brass, containing about 62 percent copper and 38 percent zinc, and was particularly suited to cold-rolling, stamping, and machining. The brass used to make the No. 17 blacksmiths' rule was a "spring" brass, containing 66 percent copper, 32 percent zinc, and 2 percent tin, which was harder to work, but was stiffer, and less subject to bending.

German (Nickel) Silver

German silver is a white alloy of copper, zinc, and nickel which was frequently used for the trimmings on premium-quality rules, due to its fine silver-like appearance, and its improved durability. This alloy originated in China, where its composition is said to have been known since time immemorial, but was only introduced to the English-speaking world in 1830, when a sample was brought to England from Germany (hence its name). For many years the alloy was very widely used for electroplated tableware and for scientific instruments, and only recently has it been largely replaced in these applications by other materials.

German silver (also known as nickel silver, Chinese white silver, Packfong, Electrum, etc.) is basically a modification of the alloy for brass, containing about 60 percent copper, 20 percent zinc, and 20 percent nickel. It has all of the workability characteristics of brass, and can be cast, rolled, and drawn. At the same time, it possesses the properties of being hard, tough, and not easily corroded. When polished it very closely resembles real silver, being only slightly more gray than that metal, and tarnishing to a light yellow, instead of an orange-black.

This alloy was used by Stanley (and most other rule makers) to trim its ivory rules, just as brass was used with the wood ones. There were some exceptions, of course; some of the smaller ivory rules were initially offered with a choice of brass or German silver trim, and a few of the least expensive in brass only, but generally speaking the rule was brass for wood, German silver for ivory. Even when the trim was of German silver, the rolls in the joints were still of brass; apparently this alloy was not available in the thick wire needed for the automatic roll-making machinery, and it was not felt to be necessary to have it specially made up.

Steel

Steel is, of course, the alloy of iron with a small amount of carbon (less than 2 percent) which is the mainstay of the industrial revolution. With its hardness and formability, this metal would have been more extensively used in rule making but for several disadvantages. Compared to brass it is too hard to scratch and stamp the markings on; if it had been used for the joint plates and tips, the scales could not have been extended to the end of the rule legs. Additionally, this same hardness would have interfered with the scraping and filing operations which followed rule assembly (see THE RULEMAKING PROCESS, later in this chapter). Equally a problem is steel's tendency to rust; such rust could probably be cleaned off exposed surfaces, but rust forming on the inner surfaces would tend to discolor the boxwood with black stains.

As a result, steel was not widely used by Stanley in making their rules. The hinge pins of the joints were steel, of course, as were the assembly pins used to fasten on edge binding and bits and to pin the buried plates of round and middle plate joints. It was never used to fasten plates on the surface of the rule; the ends of the steel pins would have been visible in contrast to the brass of the plates, and spoiled their appearance.

The only other application of steel for other than the above uses was as tips on the very cheap later rules, such as the Nos. 163 and 214T, and for the Nos. 450 and 550 counter measures.

Construction

Almost all rules were made using the same structural elements, varying only in their dimensions and in the number of "sticks" (pieces of wood or ivory) linked together by metal joints to form the rule. A rule made of two sticks with a single joint was known as a 2 fold rule, one with 4 sticks and 3 joints as a 4 fold rule, etc. Rules were classified in the Stanley catalogues by their length, number of folds, and width when folded (e.g., 2 foot, 4 fold, narrow, or 1 foot, 4 fold, broad, etc.).

The Main (Rule) Joint

The joint of a 2 fold rule, and the middle joint on 4 and 6 fold rules consisted of a circular metal disk attached to one of the two legs (the "head" stick), pinned between two thinner disks attached to the other (the "hollow" stick). The combined thickness of these three disks was the same as the thickness of the leg, and they were so arranged as to cause the two legs, when folded, to lie edge to edge. This joint was variously named the "rule", "center", "head", or "main" joint, Stanley always using the latter term (Ref. 124).

Three types of main joints were used on folding rules, all based on the same structure, but differing in their appearance, and in the method of attaching the disks to the legs. These were the round joint, the square joint, and the arch joint (see Figure 5).

In the round joint, the center disk was attached to a plate which was embedded in a slot in the end of one leg, and held in place by steel pins driven through the leg and the plate, while the two outside disks were attached to a similar plate embedded in the other leg. When assembled, only the circular disks were visible (hence the name "round"

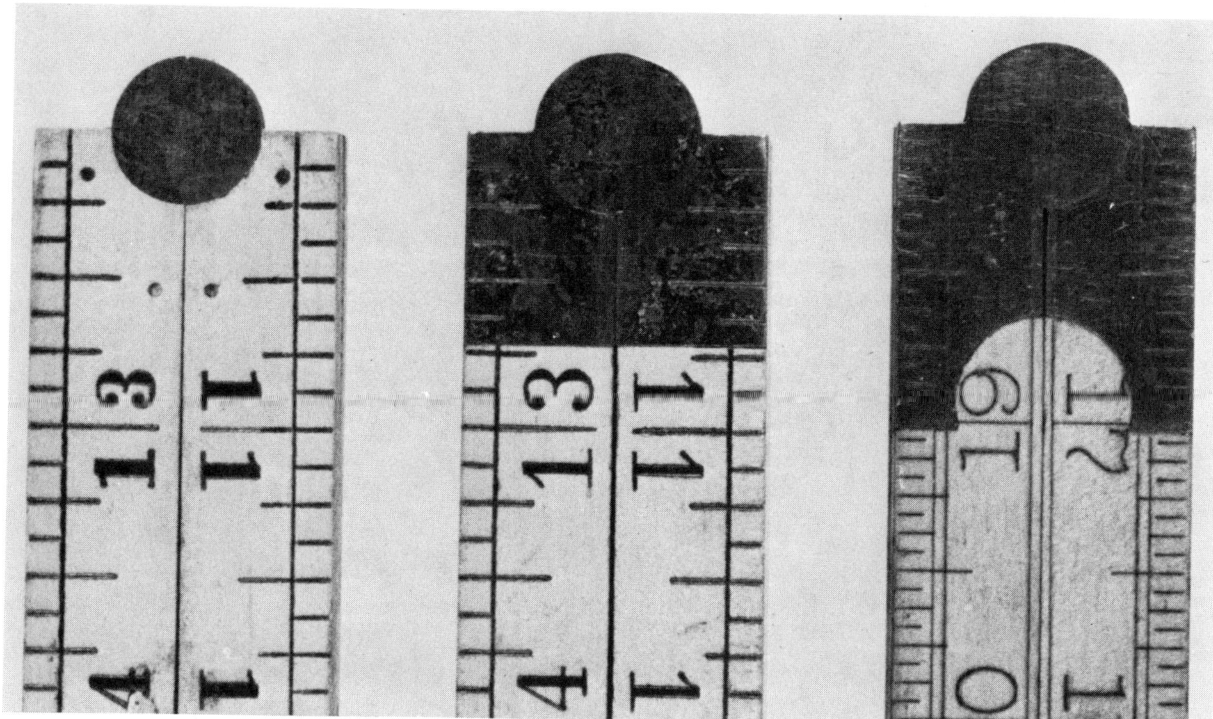

Figure 5: Round, Square, and Arch Joints *(Author's Collection)*

joint). This joint was the cheapest, weakest, and least decorative of the three types of main joint.

In the square joint, there were two plates used to attach the disks to each leg, one on each outside surface of the leg. These plates were usually "let in," so that they were flush with the surface of the leg, and held in place by brass pins driven through them and the wood between. The square joint, while more expensive than the round joint, was also much stronger, and presented a much nicer appearance.

The arch joint, dating from about 1806 (Ref. 106), was much showier than the square joint, but was structurally almost identical, and was only slightly stronger. The difference between the two lay in the shape of the plates; in the arch joint they extended farther down the leg, and were cut out in a semi-circular shape on their inside edges. Due to the extra work involved in cutting the rabbets for these plates and fitting them, the arch joint was even more expensive than the square joint, and was invariably the mark of a high quality rule.

Between 1910 (Ref. 26) and 1912 (Ref. 27) a minor change was made in the shape of the arch joint plates (see Figure 6), to reduce manufacturing cost. The original joint profile had a sharp corner at the point where the circular curve met the straight edge, and required hand work to clean out this corner when cutting the rabbets in the legs. The new shape, a modified ogee, had no such corner, and could be cut completely by machinery.

Most, but not all, Stanley multisection rules had main joints of one of these three types. There were exceptions in certain cases, however. The No. 17 blacksmiths' rule, for example, used a locking spring joint of the same kind used on wood Zig-Zag rules. The Nos. 42, 42½, and 43 ship carpenters' bevels had their tongues set into riveted friction joints. The Nos. 240 through 612 extension sticks had a sliding, instead of a rotating, joint.

The Middle (Knuckle) Joint

The other joints in 4 fold and 6 fold rules, the joints which connected the head and hollow sticks to the "tip" sticks, were similar in construction to the main joint, but had their axis of rotation parallel to the surface of the legs. As a result of this difference, the legs united by this joint, when folded, laid surface to surface instead of edge to edge.

Because the middle joints lay parallel to the surface of the leg, the joint had to be correspondingly longer (from ¼ inch, on some 1 foot, 4 fold rules, to 1½ inches, on the No. 31½ 2 fold shrinkage rule. The circular disks

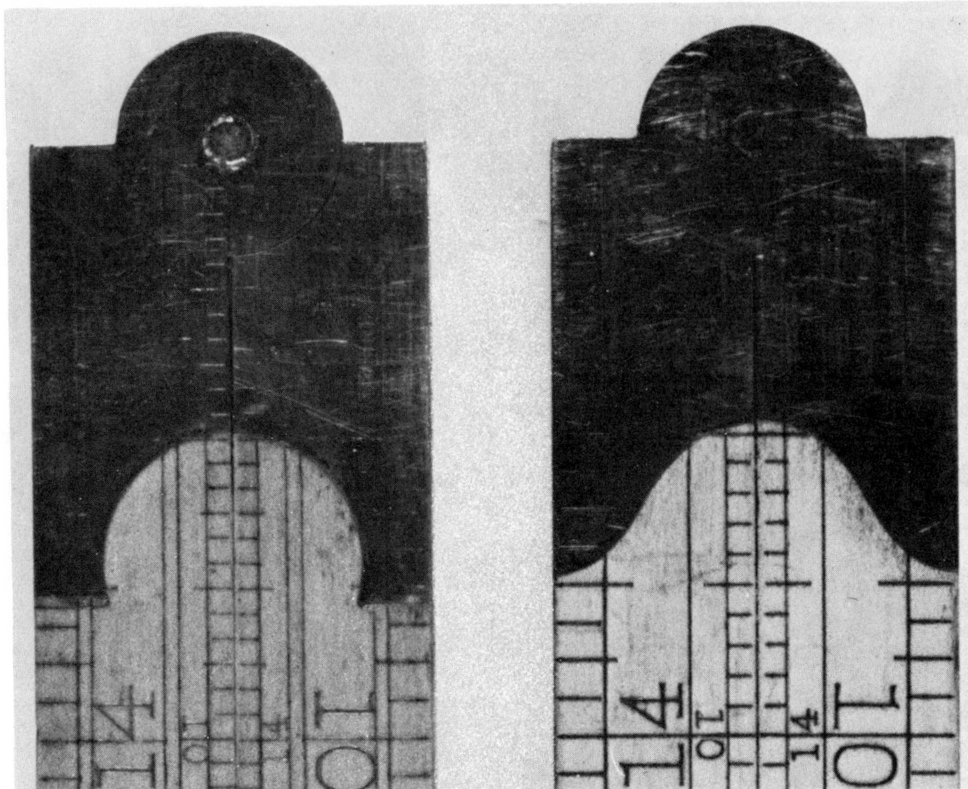

(Author's Collection)

Figure 6: Arch Joint Shapes From 1910 and Before (Left), and 1912 and After (Right)

which formed the actual pivot occupied only a small part of this distance. In order to space the disks apart, and to fill up the remaining space between them and the end of the rule, small free-turning brass rolls or collets were added to the joint during assembly.

Two types of middle joints were used in Stanley rules, depending on the quality of the rule and the strength required in the joint, known respectively as "middle plates" and "edge plates".

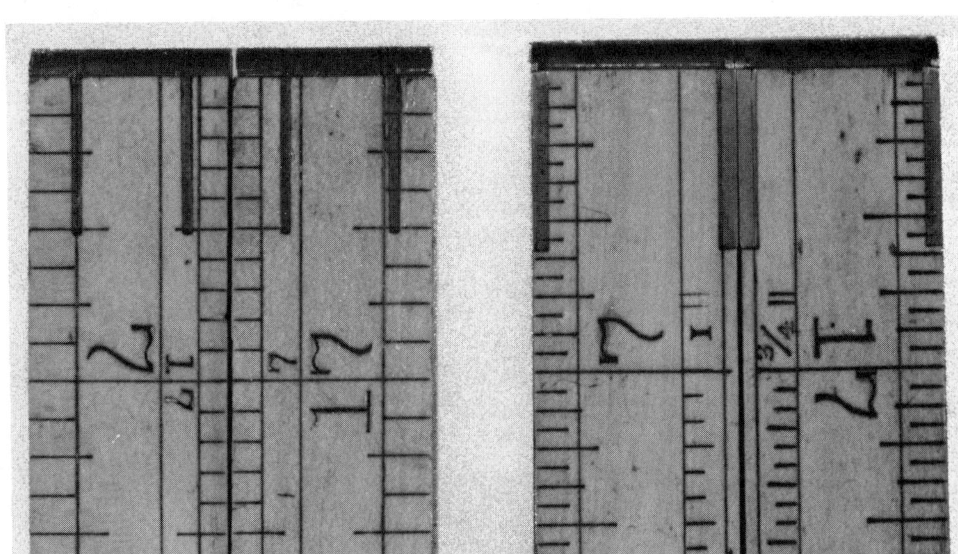

(Author's Collection)

Figure 7: Middle Plates and Edge Plates

Middle plates was the simpler form of middle joint, similar in construction to the round joint form of main joint. In middle plates the four plates, two to a leg, which formed the pivot were set in slots in the end of the leg, and held in place by steel pins driven entirely through the leg and the encased plates. Like the round joint, this was the weaker and less attractive of the two styles of plates, but was frequently used in the cheaper rules due to its lower cost.

(having the metal on the outside made this possible). Edge plates were also a better-appearing type of joint, presenting more brass to the eye, and transmitting a feeling of quality.

In certain rules the edge plate joint was made even stronger, by adding intermediate plates set into the end of the leg in the same manner as middle plates. If only a single pair of additional plates was added, as on the No. 66¼, this was called "double-plated" edge plates; if two pairs, as on the No. 31½, then "triple-plated." Double- or triple-plated edge plates were used most frequently on the longer rules, such as the No. 94, where the extra length made possible more severe wracking of the joint, and on broad, premium quality rules such as the No. 72½ and 76.

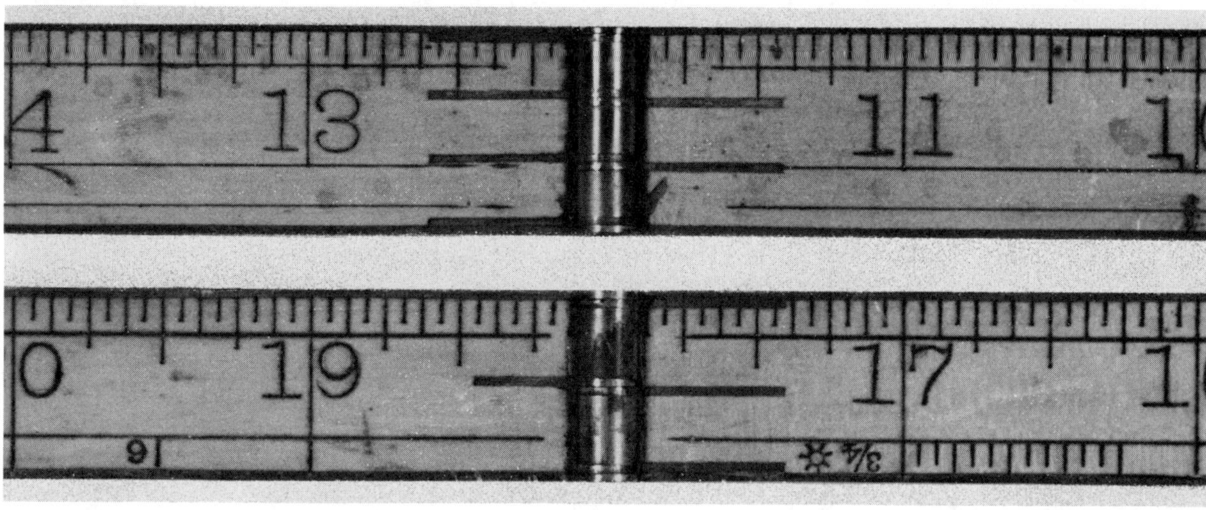

Figure 8: Double- and Triple-Plated Edge Plates

(Author's Collection)

In most rules employing middle plates each joint consisted of two pairs of plates, but there were exceptions: sometimes only a single pair of plates was used, giving a joint which was even less expensive, but also correspondingly weaker. The pattern of application of this minimal joint is not clear, but seems to have been confined to the period 1900 through 1920, and only to the lower quality 1 foot, 4 fold rules. The 1909 dealers' catalogue (Ref. 25) shows this joint fitted to a No. 69, but in 1912 (Ref. 27) this rule is again shown with two pairs of plates, but at the same time the No. 90 was shown with a single pair.

Better quality rules used edge plates, the other form of middle joint, in which the plates, instead of being set into slots cut in the end of the leg, were set flush with its outside edge, in rebates equal in depth to their thickness. This was a much stronger form of middle joint, for two reasons: the two sets of plates were further apart, giving a much better resistance to twisting, and were fastened to the leg with rivets instead of pins

The Double Arch Joint

A variation of rule construction was the so-called "double arch" joint, offered on a few of the 4 fold rules. In this scheme, the main joint was an arch joint, and arch-shaped plates were set into the outside surface of the wood of all four sticks at the middle joints to reinforce them. The inside surfaces of the sticks would be left unchanged. The middle joints could be either edge plates or middle plates.

It is not clear just how effective these extra plates were in strengthening the middle joints. In the author's experience, the typical failure mechanism for these joints was to break either when twisted or when accidentally forced past the fully open position. In either case these plates, being on only the out-

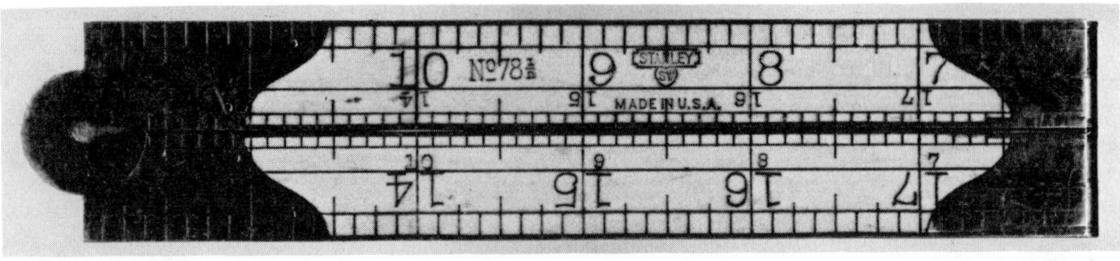

Figure 9: Rule With Double Arch Joint (Author's Collection)

side surface of the rule, could not have been of much use. They must have been primarily for decoration, and not because of any extra strength they could provide.

The double arch joint was the showiest of the four main joint designs, and was only used on the very finest rules. The extra brass or German silver presented a very fine appearance, and lent a weight to the rule which made it feel more substantial, and of better quality.

Binding

Rules were sometimes edged with thin metal strips to strengthen them, and to protect the edges of the rule from wear, a construction feature called "binding." The metal used was either brass or German silver, whichever was the appropriate trim metal for that rule, and was held in place by steel pins driven through it into the body of the rule. Two forms of binding were offered, half binding and full binding.

On half bound rules only the outside edges were bound, the edges which were exposed when the rule was fully folded. These were the edges which would suffer wear while in the pocket or tool chest, and needed the most protection.

On full bound rules, all edges of the rule, both inside and out, were edged with metal. This was the most elaborate form of binding, offering the most protection, but also having the highest cost.

Whether a rule was bound was a function of the rule number. Full bound 6 fold, 4 fold, and 2 fold rules were all offered, and half bound 4 fold and 2 fold rules as well (although in this latter case only for a short time; 2 fold half bound rules were only available from 1855 to 1859, probably due to problems with warping).

Bound 4 fold rules always had the edge plate type of middle joint, and, in such cases,

Figure 10: Unbound, Half Bound, and Full Bound Rules (Author's Collection)

the binding plates also served as the plates of the joint. This was true whether the rule was half bound or full bound.

Binding had other disadvantages in addition to cost. A half or full bound rule was significantly heavier than its unbound equivalent, enough so that one could feel the difference. Additionally, binding a rule prevented the use of the edge for an additional scale; the bound carpenters' sliding rules, for example, do not have the 100ths of a foot scale on their outside edge which was a feature of the unbound ones.

Alignment (Closing) Pins and Bits

A common problem with 2 fold and 4 fold rules was the bending and breaking of the joints which could occur if the joint was bent past the fully open position, or the leg was inadvertently bent sideways when shut. Not much could be done to prevent the first type of abuse, but the second could be largely prevented by alignment pins. These alignment pins (or "closing pins," as Stanley referred to them) were perpendicular steel pins set in the inside surface or edge of the rule, which mated with corresponding holes in the opposite surface when the rule was closed. They prevented the legs containing the pin and socket from sliding across one another when shut, and thus prevented damage to the joint connecting them.

The pin and hole protecting the main joint were located on the inside edge of the

(Author's Collection)

Figure 11: Alignment Pins on a 4 Fold Rule

rule near the tips of the legs (if a 2 fold rule), or near the middle joints (if a 4 fold rule). High quality 2 foot, 2 fold rules would sometimes have two sets of pins and holes, the second set being located midway between the joint and the tips.

Four fold rules also had alignment pins to protect the middle joints. These were located on the inside surfaces of the rule near the tips and the main joint, one pin and hole set to protect each joint.

Some carpenters must have felt that the alignment pins were in the way, by preventing them from laying the edge or surface of the rule flush on the work when measuring or marking. A number of examples have been found in which the surface pins, and sometimes the edge pins as well, have been carefully removed. Apparently they were more willing to put up with the possibility of damaging the rule joints than with having the pins in the way.

The alignment holes in the edge of the rule were rather delicate, and it is common to find examples where the pin has been pulled sideways from the hole, tearing the wood and damaging the edge. Bound rules, and unbound 4 fold rules with edge plates did not have this problem; on such rules the hole would naturally fall on a part of the edge protected by the binding or plate and thus the pin could not tear out. Unbound and half bound 2 fold rules, and unbound 4 fold rules with middle plates had no such protection, and were often damaged.

On a few boxwood rules of these last three types it was sometimes the practice to install "bits" to protect the edge from this pin damage. These bits, substitutes for the binding or joint plates which protected the more elaborate rules, were rectangular brass plates, perhaps ½ inch long, set flush and pinned into the edge of the rule and drilled to receive the alignment pin(s).

The selection of which rules should have bits seems to have been somewhat arbitrary. The No. 4 obviously had them because of its extraordinary thinness, but why the No. 27 should have them, and not the No. 26, for instance, is entirely unclear.

At various times 15 different Stanley rules were equipped with bits. Eight were 2 foot, 2 fold rules, the Nos. 2, 4, 6, 12, 14 (probably), 19, 22, and 27; four were 2 foot, 4 fold rules, the Nos. 59, 71, 74, and 77; three were 2 fold caliper rules, the Nos. 13, 13½, and 36½. Only about half of these rules were always equipped with bits, the others having them added or deleted at various times during their life. The caliper rules were not equipped with bits until 1912. The last 2 foot, 4 fold rule with bits was discontinued in 1910, the last 2 fold rule in 1917, and the last caliper rule in 1932.

Tips

The tips of the rules were U-shaped pieces of metal which wrapped around the two ends of the rule and were set flush with its surface. The metal used was the same metal which formed the joint plates of the rule; this was usually brass for wood rules, and German silver for ivory ones. A few rules were equipped with steel tips, either plain or brass-plated, but only the least expensive, and those only just prior to the end of 4 fold rule manufacture.

Two types of tips were offered at various times on the long, nonfolding measuring sticks, U-shaped brass tips similar to those used on the folding rules, and a 2 piece sandwich design which set and riveted a separate plate into each of the two opposing surfaces, but left the end of the stick exposed. Judging from examples of both types examined, the two plate form is an earlier variation, and was probably discontinued before 1910.

Slides and Calipers

A number of Stanley rules were equipped with "slides" set into the surface of one of their legs. These slides were made of thin brass, with a hat-shaped cross section, and ran in an undercut groove in the wood or ivory of the rule leg; thus they could slide back and forth freely in the groove, but could not be removed except by sliding them out.

Two types of slides were made, ordinary and caliper.

The simplest form of ordinary slide was the so-called "plain" or extension slide. This was a slide graduated in 8ths of inches, and was used to extend the length of the rule when taking inside measurements. Three rules had plain slides: the No. 26 2 fold rule, the No. 212 hatters' rule, and the No. 83 rules made after 1874.

The other form of slide was the Gunter's slide, a primitive type of slide rule. Gunter's slides were standard on the carpenters' and engineers' sliding rules, and also on the No. 83 4 fold rules prior to 1877. On this slide, both edges of its surface, and both edges of the groove, were engraved with logarithmic

(Author's Collection)

Figure 12: U-Shaped and "Sandwich" Tips on Stanley Yard Sticks

scales; by sliding it back and forth in the rule leg it could perform all ordinary computations dealing with multiplication, division, and ratios. On some of the later rules with Gunter's slides, the back of the slide was graduated in 8ths of inches, thus allowing it to function as an extension slide as well.

Caliper slides were used for measuring the outside dimensions of small objects (typically up to 2-3 inches), and were equipped with a jaw on the end which overhung the tip of the leg in which it ran. The length of the slide was graduated in inches, beginning at the jaw; by placing the object between the jaw and the tip of the leg and bringing the jaw down agains it, its dimension could be read on the graduations on the slide.

Stanley manufactured nineteen different rules with caliper slides between 1855 and 1975. Eight of these were 4 fold rules, the Nos. 3, 32, 32½, 39, 40, 62C, 76C, 83C; nine were 2 fold, the Nos. 13, 13½, 14, 35, 36, 36½, 37, 38, and 40½; and five were nonfolding, the Nos. 23, 24, 210, 136 and 136½. Of these, six were ivory, the Nos. 24, and 37 through 40½, and the rest boxwood.

Up through 1932 the standard caliper slide arrangement was left hand, that is, the caliper was set in the leg of the rule and graduated in such a way that when measuring across the top of an object with the caliper jaw pointing down, the caliper would be held in the left hand and the object in the right. This was the customary arrangement in the United States, in contrast to England and continental Europe, where right hand calipers were the norm (see Figure 13).

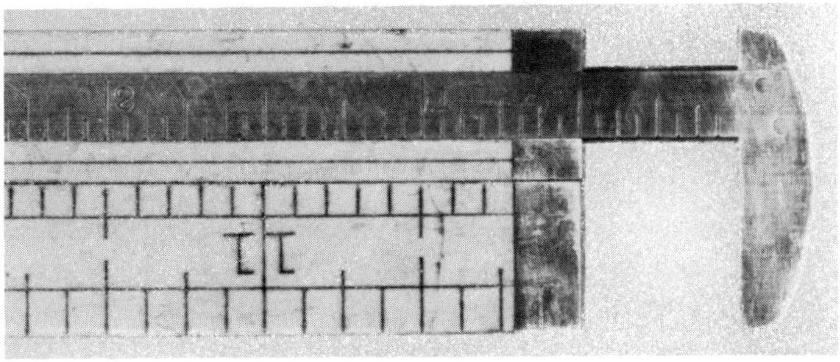

Figure 13: Left Hand (Left) and Right Hand (Right) Calipers *(Author's Collection)*

In 1934 (Ref. 44) Stanley changed from left hand calipers standard to right hand calipers standard. By that time only six caliper rules were still in production, the Nos. 13½, 32, 32½, 36, 36½, and 136. The No. 136 already had a right hand caliper; the other five, all left hand up until 1932 (Ref. 43), were switched over at this time.

The Rulemaking Process

Prior to about 1850, rule making was a craft, conducted in small 5 to 10 man shops, and the processes and techniques were passed from master and journeyman directly to apprentice. The large rulemaking firms which introduced mechanization after that date usually made their own specialized machines, and chose to protect their methods by secrecy. As a result, very little published material exists describing the process of making folding rules.

Only one short article (Ref. 124) has been found describing rule making at the Rule & Level. The description which follows is based largely on rule making at D.H. Stephens & Co. (Refs. 109, 116, and 145), and at John Rabone & Sons, Birmingham, England (Ref. 106), but is probably reasonably accurate, the basic steps in the process were common to all makers, the differences among them being primarily in the degree of mechanization and in minor manufacturing details.

The Raw Materials

The process of manufacturing a rule actually began several years prior to its final assembly, with the acquisition and preparation for use of the boxwood and ivory.

The boxwood was initially received (in the form of logs) at the Stanley seasoning and storage yard (called "The Farm" by Stanley employees) outside New Britain. In its heyday, Stanley obtained its boxwood by the half-shipload directly from Turkey or Venezuela (Ref. 128), transshipping it by train from the dock to the yard. At "The Farm" the logs, from 8 to 10 feet long, would be split into either halves or quarters (depending on diameter) and stored under roof in open air-drying sheds.

The ivory was acquired in the form of blocks from a New York City firm which imported it from Africa (sometimes directly, other times via London or India). Elephant ivory was the type used, the tusks of the other mammals being too small to yield the 6 inch lengths needed for 2 foot rules.

When the boxwood was fully seasoned (a matter of several years) the logs were then ready to be made into sticks. As the first steps in this process they were "blocked up" (that is, cut into blocks of the proper length), and then sawn and resawn lengthwise into rough sticks slightly wider and thicker than the desired final size. Great care was taken during these sawing operations to ensure that the cuts were made parallel to the grain of the wood. The rough sticks were then "dressed off," milled straight and smooth on all four surfaces, and sized to their final width and thickness. This was done by different means at various times in Stanley's history: at first by the traditional method of hand planing with special high angle planes, then later by methods more adapted to large volume manufacture, such as drum sanding and milling machines.

The ivory was also allowed to season before use, first in block form, as it was received from the dealer, and then again later after it had been sawn into sticks. It was even more important to season ivory than boxwood; it is characteristic of ivory that it can shrink as much as 10 percent as it dries from the green (freshly removed from the elephant) to the dry state. Any rule made from incompletely seasoned ivory would shrink to be too short and/or warp, and if bound, would probably crack from the strain.

When fully seasoned, the ivory pieces would also be "blocked up" by sawing and resawing. This was done on special fast-running, fine toothed saws, fed very slowly to prevent scorching and cracking.

The second episode of drying the ivory came next. The sawn sticks would be clamped in a press to prevent warping, and placed in a warm, dry place until finally completely dry. At Stephens & Co. they were placed in a glassed-in cupola, so that the sunlight could help retard any tendency to yellow while drying (Ref. 145); presumably Stanley used a similar method.

The metal to form the various parts of the rules was received in roll form: flat brass sheet for the various plates and trim pieces, large diameter brass wire for the rolls in the joints, and smaller diameter steel and brass wire for the joint and assembly pins.

Manufacturing and Assembly

There were three distinct steps in rule making: joint making, framing, and finishing. Joint making was concerned with the shaping of the different metal parts for the joints and putting together the joint assemblies ready for use. Framing was the process of fitting the assembled joints and trim pieces to the sticks to form a complete rule. Finishing dealt with applying the graduations, figures, and other markings, and applying a protective finish to the surface.

Joint Making

Insofar as was possible, the metal parts for the joints were cut and shaped on automatic machinery. The hinge plates, tips, binding strips, and bit plates were die cut from sheet brass, and the tips folded and squared in a press. The rolls for joints were turned and drilled on automatic lathes (screw machines). The steel hinge pins, and the steel and brass assembly pins were cut from wire on still other machines.

After being cut to shape these metal parts had to be "scraped" to eliminate any burrs left by the presses and lathes, and bring them to their final shape. Any holes required were made at this time, and the large holes which held the pivot pins were countersunk with the peculiar 5-sided shape required to match their heads (these 5-sided heads guaranteed that the pin would be locked to the outside plates of the joint, and rotate only with respect to the inside plates).

After the parts were prepared, they were "driven up," that is, assembled into finished joints. The various plates and rolls would be placed in a jig which would hold and align them, and then clamped while the pin was inserted and then headed on the other side.

Joints for wood rules would be assembled from all brass components. Those for ivory rules would use mostly German silver parts, except for the rolls which spaced the plates apart; for some reason these were always brass irrespective of the type of trim.

Framing

The first step in framing was to cut the dressed off sticks to exact length, and then to cut the various recesses in their surfaces and ends for the trim and type of joint they would receive. Quarter circular cuts were made for the rolls which spaced the joint plates apart (on both ends of the head and hollow sticks, on only one end of the tip sticks). If the rule was to have a round joint or middle plates, slots were cut to accommodate the joint plates; if a square, arch, or double arch joint, or edge plates, shallow rabbets would be cut in the stick surface to house them. The tips (and bits, if installed) would also require rabbets/dadoes. The slot for a slide or caliper would not be cut at this

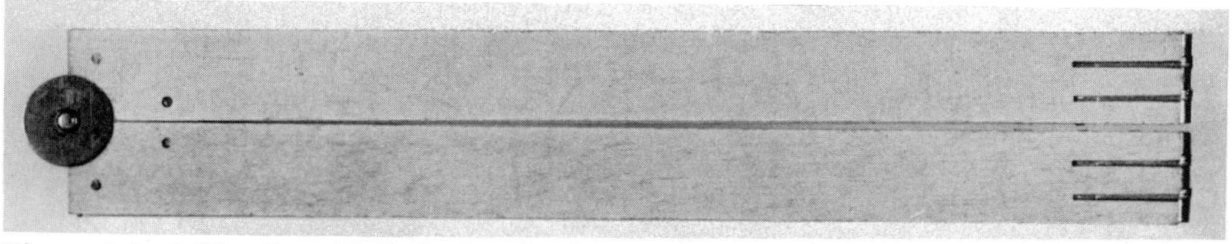

Figure 14: A Number 68 Rule Ready for Graduation *(Charles and Walter Jacob Collection)*

time, however; this was postponed until after the tips were installed, so that both could be grooved in one operation.

Assembly was next. The trim (the tips, bits, and binding) was attached to the separate sticks first, riveted to the surface by pins driven into predrilled holes in the wood or ivory, and then attached. The first joint fitted was the main joint, connecting the head and hollow sticks. After it had been pinned/riveted in place, then the two middle joints were fitted, attaching the two tip sticks.

At this point, the length of the rule was adjusted to its exact final value. This was done by sanding the tips; the combined length of the sticks and the thickness of the brass tips was such that the newly assembled rule came out slightly oversize; removing a few thousandths brought it down to exactly the correct length (the resultant thinning of the tip plates where they wrap around the end of the stick is usually visible if looked for).

After the length of the rule had been corrected, the various trim and hinge plates were brought flush with the surface of the sticks. Being (by design) a little thicker than the rabbets prepared for them, they protruded slightly above the surface of the sticks, and had to be scraped/sanded flush. Wherever possible, such as on the outside face and edge of the rule, this was done by machine; on the inside face and edge, where the joint protruded above the surface, a combination of machine and hand work was required.

The final step in framing was to drill for and install the closing pin(s). They would have interfered with the scraping/fling process, so they were not put on until the rule was otherwise ready for graduation. Only the edge pin(s) were put in at this time. The surface pins had to wait until even later, after the graduating process was complete; they would have prevented the rule from lying flat when graduating the outside, and blocked the dividing knife when scribing the inside.

This was the final step in assembly. After it was complete the rule was ready to be turned over to the finishers to be graduated and figures (see Figure 14).

Finishing

Finishing was an area in rule making where it was considered suitable (and economical) to hire women. No great physical strength was required, only a good eye and a steady hand, patience and concentration, and a willingness to work for low wages.

The first step was to seal the surface of the blank rule by applying a coat of clear finish. The basic graduation process (until the introduction of printed graduations and figures after 1900) was to scratch or stamp fine lines into the surface of the rule and then fill them with a black paste to make them visible. By sealing the rule surface prior to this operation the paste was prevented from staining the wood when being rubbed into the lines.

Initially the graduation lines were scribed by hand as follows:

Straight scales and rules are usually divided by placing the article to be divided and the original pattern side by side, then passing a straight-edge with a shoulder fixed at right angles to serve as a guide along the original, and pausing at each division; then a corres-

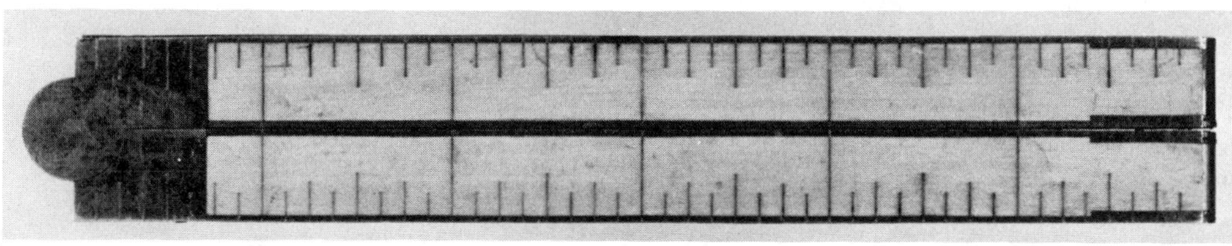

Figure 15: A Number 62 Rule Graduated With 8ths, Ready to Be Graduated Next With 10ths and 12ths *(Author's Collection)*

ponding line is made on the copy by the dividing knife.

This was a more rapid process than might be imagined. A skilled worker could do an entire scale in a very short time (Rabone states that the 1 foot Gunter's scale of the carpenters' sliding rule, comprising 540 lines, could be cut by hand in 10 minutes).

The introduction of factory methods to rule making included the mechanization of graduation as well, however. Machines were developed where 50 or 100 rules at a time could be clamped to a large table, each below a separate dividing knife; these ganged knives were controlled by a complicated mechanism which moved them in unison to simultaneously scribe each line in succession on all the rules at once. Some of these machines used a lead screw to advance the knives each time and a cam to control alternating line length; others controlled them via a pantograph-type linkage which traced a large scale master pattern.

Slides and calipers were graduated separately from the rules to which they were fitted. At first this was done by scribing, in the same way that the wood was marked, but by about 1900 methods had been developed to stamp the graduations into the metal with a rolling die, at a great saving in time.

Prior to separating the body and slide to graduate them, they were marked, so that each slide could be reunited later with the body it had originally been fitted to. This was done with a number stamped on the back of the slide and an identical number stamped on the inside of the groove. Each rule/slide in a lot would have a different number, thus making identification simple (it is interesting to observe that these numbers all seem to fall in the range 1 to 50; this may indicate that the usual lot size for graduation was 50, and thus, by implication, that the graduating machines could only handle 50 rules at a time).

After the graduations had been scribed, the figures were then stamped into the rule. Special stamps were used, with a finer edge than those for steel, which would cut the wood fibers instead of crushing them. Ivory, being so hard and brittle, was steamed to soften it temporarily for stamping.

Like most other phases of rule making, figuring was eventually automated, with stamping and figuring machines being developed for this task.

The final step in marking the rule was to apply the rule number and the inside and outside trademarks. This was also done by stamping, with special stamps each of which could mark one or more lines of text at a time. Sometimes the trademark would be omitted, so that a finished rule could be sold to a wholesaler or retailer who wished to market rules under their own name, or so that it could be transferred to another maker as part of the process whereby the rule makers' association adjusted production volumes to sales (Ref. 142).

The lines created by this scribing and stamping were then filled with a black paste. Linseed oil mixed with lampblack was used at the Chapin-Stephens Co.; Stanley may have used this also, or some other type of paint or wax instead. It was applied liberally with a brush, and then the surplus wiped off with a soft cloth; the residue left in the graduations and figures rendered them both visible and durable.

After the blacking had dried, a second, and possibly a third coat of clear finish was applied to the rule. This protected the surface and the markings from damage and wear, and gave the rule a smooth, attractive finish. The practice of polishing the final coat, to make it extra smooth, was apparently discontinued about the time of the First World War; rules made during the 1920's have nowhere near the fine surface finish seen in the earlier product.

The last step prior to final inspection and shipment was the installation of the inside closing pins, which, as mentioned earlier, had been postponed until after graduation. Once they were in place, the rule was complete and ready for use.

CHAPTER 3 Graduations and Markings

Trademarks and Rule Numbers

A. Stanley & Co. Markings (1854-1857)

The boxwood rules manufactured by A. Stanley & Co. were marked with the rule number, inside and out, and stamped on the inside, near the main joint, with the phrases: A. STANLEY & Co, NEW BRITAIN CONN, and WARRANTED BOXWOOD. This was usually stamped in two rows, the first two phrases on the lower edge of one leg, in $^{3}/_{32}$ inch letters, and the third below, on the upper edge of the other leg, in $^{1}/_{8}$ inch letters (see Figure 16).

Figure 16: The A. Stanley Marking Stamped on a Boxwood Rule *(Paul Kebabian Collection)*

Ivory rules seem to have been marked with the rule number on the outside only, and to have usually had the inside phrases stamped in two lines, both on the upper leg near the main joint (see Figure 17). The reference to boxwood was, of course, omitted.

Figure 17: The A. Stanley Markings on an Ivory Rule *(Bud Steere Collection)*

This marking scheme was not uniformly adhered to, and numerous variations exist. The three most common are boxwood rules where the company name and factory location are in two lines instead of one, or where the markings are all on one leg, and both boxwood and ivory rules which have no rule number markings at all.

Stanley Rule & Level Co. Markings (1858–1921)

At or shortly after the merger with Hall & Knapp the trademark stamped on the rules was changed to reflect the new company name. The rule number continued to be stamped, inside and out, as before, but the maker's name was changed to read: STANLEY RULE & LEVEL Co., NEW BRITAIN, CONN. On boxwood rules this was stamped, as before, on the inside near the main joint, in three lines of $^{1}/_{16}$ inch letters. Sometimes all three lines were on the upper leg, sometimes they were divided, with two lines on the upper, and one on the lower.

Usually, but not always, the phrase WARRANTED BOXWOOD was stamped either to the right of or, more commonly, below the maker's name, also in 1/16 inch letters. Shortly after the adoption of this marking scheme the letters U.S.A., identifying the country of origin, were added, either to the right of the state name, or on a fourth line below it (see Figure 18).

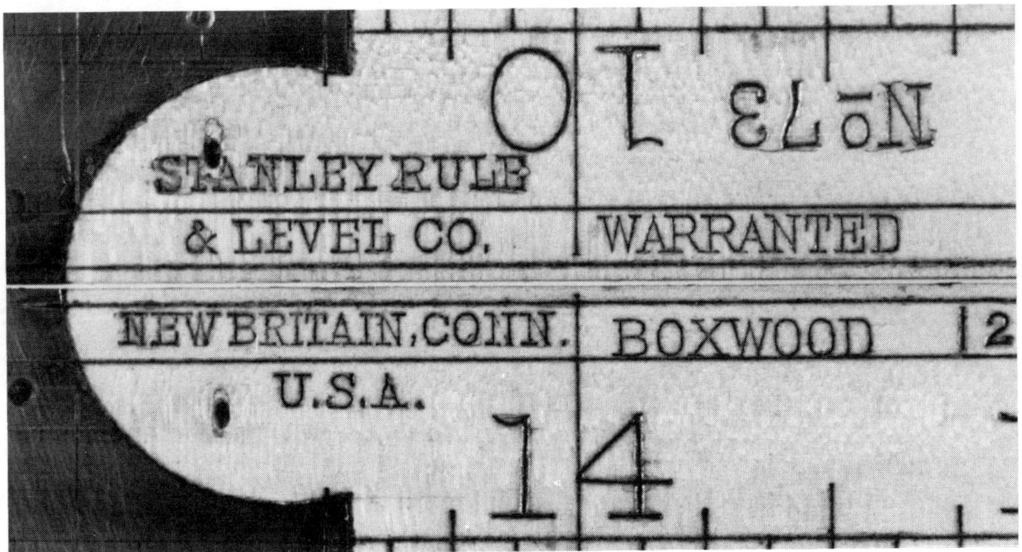

(Author's Collection)

Figure 18: Stanley Rule & Level Company Markings on a Boxwood Rule (Typical)

At the same time, the inside stamp on the ivory rules was changed to read STANLEY or (on two lines) STANLEY, NEW BRITAIN, CONN (examples of both have been noted). There was a tendency to use simpler stamps on ivory rules, due to their brittleness and the danger of breakage. The rule number was stamped either on the inside or (more commonly) the outside, but not both.

On 1 foot, 4 fold ivory rules the name STANLEY was stamped on the outside with the rule number, or equally often, the rule number alone.

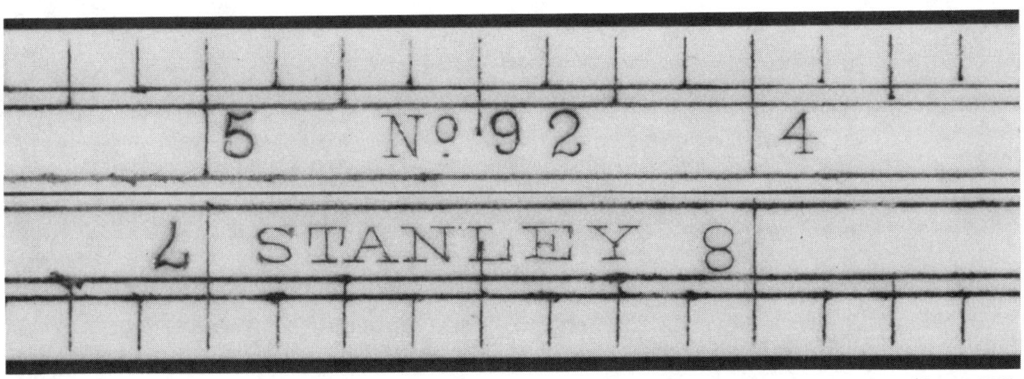

(James Hill Collection)

Figure 19: Stanley Name and Rule Number on a 1 Foot, 4 Fold Ivory Rule

Sometime between 1879 and 1898, the marking on the outside of the boxwood rules was changed to include the name STANLEY in 5/32 inch letters, to the right of the rule number (see Figure 20). At about the same time the arrangement of the markings on the inside was standardized, with the company name on the upper leg in two lines, the factory location in two lines on the lower, and the warranty split between them in two lines to the right, and the letter size was changed to 3/32 inch.

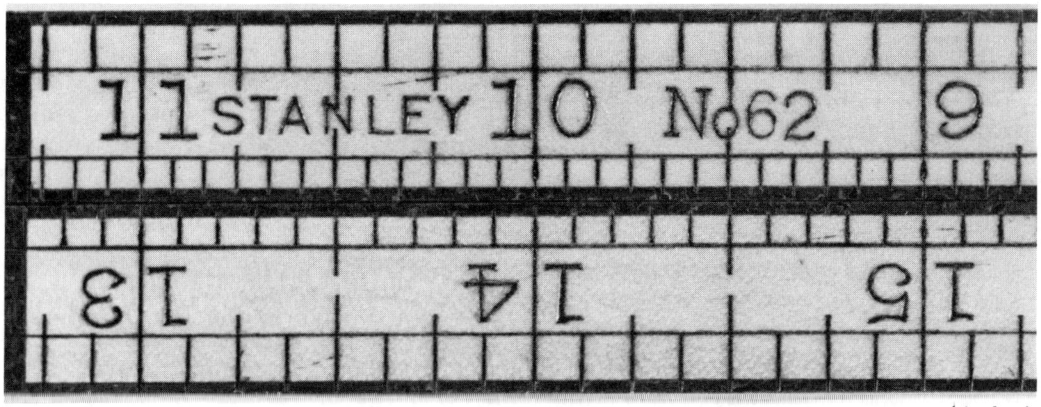

(Author's Collection)

Figure 20: The Stanley Name and Rule Number on the Outside of a Rule, 1879–1921

The markings on the ivory rules were also changed at some point in the Stanley Rule & Level Co. period. The Stanley name was moved to the outside, and changed to read STANLEY RULE & LEVEL CO. NEW BRITAIN, CONN. USA. This was stamped in a single line down the middle of the lower leg (see Figure 21), with the rule number stamped above it on the upper leg. We do not know at what point during 1858–1921 this occurred; the line of ivory rules was not dynamic enough, and there are too few samples available to bracket the changeover.

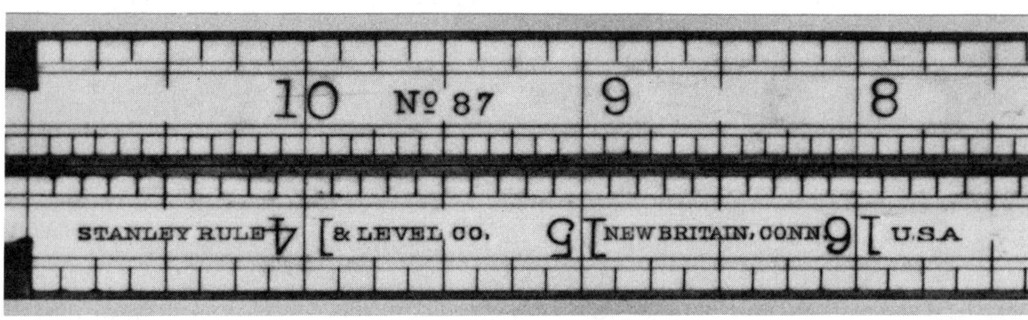

(James Hill Collection)

Figure 21: Stanley Rule & Level Co. Markings (Single Line) on the Outside of an Ivory Rule

"SWEETHEART" Markings (1922–1933)

After the merger with the Stanley Works in 1920, the markings on the rules were changed once again. The company name and location were deleted from the inside, leaving only the boxwood warranty in the same location as before. The markings on the outside were changed to be the Sweetheart trademark, the rule number, typically in ¼ inch figures, and the phrase, MADE IN USA, in ⅛ inch letters (see Figure 22).

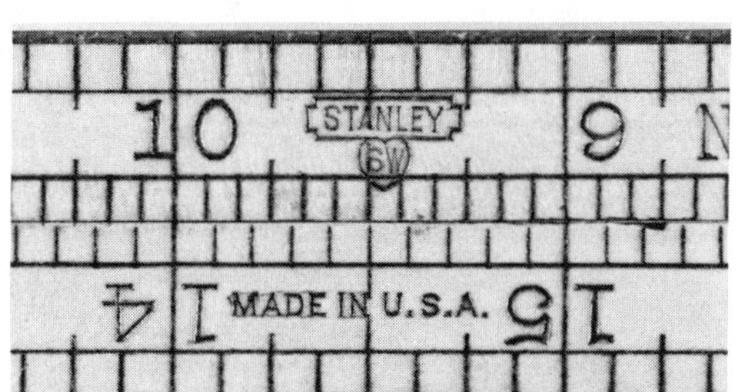

(Author's Collection)

Figure 22: The Sweetheart Trademark on a Stanley Rule

"MADE IN USA" Markings (1934–Present)

The final change in the markings on Stanley rules came in 1934, when the company abandoned the Sweetheart trademark in favor of the simpler rectangular "Made In USA" trademark adopted in that year. The trademark was the only thing changed; the other markings on the rules remained the same (see Figure 23). It is not possible to generalize about the size of this trademark and the letters and figures used for the rule number, etc., during this period. There apparently was considerable variation, depending on whether the markings were incised or printed, on the width of the rule, etc.

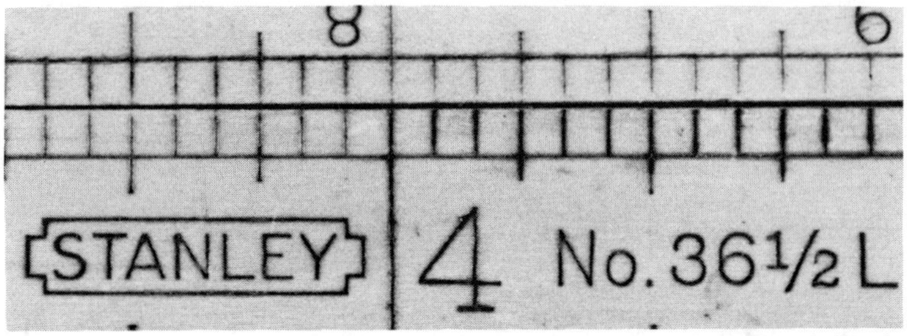

(Author's Collection)

Figure 23: The "Made In USA" Trademark on a Stanley Rule

EMBELLISHMENT LINES

Embellishment lines are the longitudinal lines, parallel to the rule edge, marked on all but the least expensive rules. These lines were marked on both the edge of the rule and the face, each pattern of rule having its own combination of lines.

Face embellishment lines were marked either singly or in pairs, depending on the style of the particular rule, and were usually placed between the figures and the edge, crossing the graduation lines. Some rules also had additional lines above the figures as well.

The original purpose of the face embellishment lines was as an aid to the early rule makers during graduation. When marking a rule by hand it was very helpful to the workman scribing the graduations to have a line marked on the rule surface to indicate the correct length of each one. The 16ths graduations could be started a little short of this line, the 8ths graduations just at it, and the 4ths graduations slightly beyond it, etc.

Even after the advent of machine graduation of rules in the late 1800s, the face embellishment lines were retained on all but the very cheapest rules. Just as they had guided the workers in hand marking the graduations correctly, so they also made it easier for the user to read them, distinguishing the fine and coarse divisions from one another. Additionally, they also improved the appearance of a rule, "framing" the graduations and figures, and giving them an attractive finished look.

The simplest scheme of face embellishment line was a single line crossing the graduations even with the tips of the 8ths graduations. Elaborate ones (reserved for premium rules) could have as many as four

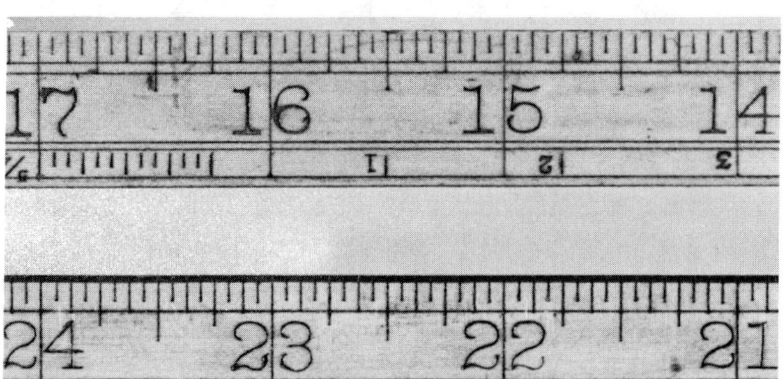

(Author's Collection)

Figure 24: Simple and Elaborate Face Embellishment Schemes

embellishment lines per scale, two in the region of the graduations, and two more, purely decorative, very close together on the other side of the figures, at the point where the inch graduation lines ended. Some of the early rules even had as many as seven, including a line at the outer edge of the graduations.

Edge embellishment lines were much simpler, consisting of a single line or pair of lines scribed the length of the edge, dividing it into either two or three equal width areas. They were purely decorative, and were omitted if the edge was bound, or was graduated (with a 100ths of a foot scale, the only scale ever marked on the edge of a Stanley rule).

GRADUATIONS

Stanley rules were graduated with a wide variety of different scales, the particular combination on any rule being a function of its intended use and whether it was an ordinary or a quality product. Broadly speaking, three classes of graduations were marked on the rules: linear, 1:1 (full size) scales, linear scales which were not 1:1, and nonlinear scales.

Linear, 1:1 Scales

The most common of these were those which were "true measure" or "1:1" scales, scales which were intended to measure linear distances and indicate the result in commonly accepted units of uniform size, such as inches, millimeters, yards, etc. The fractions of inches scales on the vast majority of the Stanley rules are an example of this class. Scales of 8ths, 10ths, 12ths, and 16ths of inches were all common. Scales finer than this are rare: presumably because the wood fibers between adjacent lines would be so short that they would tend to separate from the surface with only slight wear. The only Stanley rules where the wood was marked with scales finer than 16ths of inches were the metric rules (1 millimeter = approximately 1/25 of an inch) and the No. 036 combination rule.

English Graduations

8ths & 16ths of Inches

Although these graduations usually went together, this was not universally the case. Sometimes a scale would be graduated in 8ths only.

Each line in the 8ths/16ths scale was a different length, as a function of the fineness of the division it represented. The 16ths markings were the shortest, not quite reaching the embellishment line. The 8ths markings were longer, ending at the line, and the 4ths markings even longer, extending past it (to the second embellishment line, if such was present). The half inch markings in their turn were correspondingly longer still, and the inch marks, of course, extended all the way past the figures (to the embellishment lines beyond them, if any).

The 8ths and 16ths scales were always marked adjacent to the outside edge (when folded) of 2 fold, 4 fold, and 6 fold rules. This allowed these commonly referenced scales to be used free of interference from the alignment pins located on the inside edge. On 4 fold rules the usual arrangement was to mark the 8ths scale on the outside (when folded), and the 16ths scale on the inside. Presumably this was done so that when the 16ths scale was used, the rule could be laid flat with this scale uppermost without similar interference from the inside surface alignment pins. This arrangement also had the advantage of protecting the 16ths scale from dirt and pocket wear when the rule was closed.

10ths & 12ths of Inches

It was not possible to mark the 10ths and 12ths scales with the same hierarchical arrangement that had been applied to the 8ths and 16ths scales, which had the property that each division was 1/2 of the previous one. Only the whole inch graduations, and those subdivisions which coincided with 4ths of inches were emphasized (5/10 ths on the 10ths scale, and 3/12 ths, 6/12 ths, and 9/12 ths on the 12ths). All other subidivisons were of the same length, reaching exactly to the embellishment line.

As far as can be determined from catalogues, etc., up until 1874, only a few rules were available with 10ths of inches scales, and when such a scale was a feature it was marked on the edge of the rule. Beginning in that year (Ref. 10), however, this scale became a feature of a great many more rules, and, equally of interest, from then on was always depicted on the surface of the rule in new catalogue illustrations. This change was probably the result of the introduction at this time of new graduating machines capable of marking 10ths scales. These machines could typically only mark graduations on the surface of the rule (hence the change in scale location) but reduced the cost of applying this scale to the point where it could be much more widely offered.

A similar situation prevailed about 1879 with regard to the 12ths of inches scales. Beginning in that year (Ref. 12), 12ths scales, which had previously been relatively rare, began to be offered on a great many rules. It is probable that this was due to the same cause as the suddenly wide availability of 10ths scales three years before: A new graduating machine had been introduced which could mark this scale, or, more likely, the machine introduced in 1874 had been modified to mark 12ths scales as well.

24ths of Inches

The 24ths of inches scale was one of the two exceptions to what was apparently a Stanley policy to never graduate the wooden body of a rule with any scale finer than 16ths of inches. Presumably when scales finer than that were marked on boxwood the severed wood fibers between successive lines were so short that they were not firmly enough attached to the adjacent fibers and could too easily be abraded off.

The reason for this exception is that this scale was a recognized feature of the No. 036 combination rule which Stanley had acquired as part of the residue of the defunct Chapin-Stephens Co. in 1929. This rule was the one rule which was continued in production after that purchase, and apparently it was decided to market it unchanged, even though one of its scales was graduated more finely than Stanley ordinarily would permit.

Only every third line in this scale (the lines coinciding with the 8ths of inches) was emphasized, all the others ending at the first embellishment line; the 8ths lines extended slightly beyond it, the 4th lines even further, and the half inch and inch lines all the way beyond the figures to the second embellishment line.

100ths of Feet

This scale was not generally offered, as its applications were rather specialized (it was useful in engineering calculations and measurements dealing with distances expressed entirely in feet and decimals of feet (e.g., 2.16 feet, etc.). In ordinary carpentry, where measurements were usually given in feet and inches, it was not applicable, and would not have justified the extra cost of marking it.

This scale, when present, was always marked on the outside edge of the rule (presumably by hand). It had no embellishment lines, but simply consisted of lines halfway across the edge at each 100th of a foot, with the 20ths lines slightly longer, and the 10ths lines going completely past the figures. It was figured in 100ths, with a figure (10, 20, 30, etc.) each 10th of a foot.

Fractions of a Yard

Fabric for sewing, drapery, and upholstery purposes is usually priced and sold by the yard, and the traditional measuring instrument for this purpose is a 3 foot stick marked on one side with a scale of fractions of a yard. Stanley manufactured several "yard sticks" graduated in this fashion.

The usual arrangement was to have one surface of the rule/stick devoted solely to this scale, with markings at the 1/32, 1/16, 1/8, 1/4, 3/8, 1/2, and 3/4 yard points. This system was not uniformly adhered to, however, and variations, such as the addition of 5/8 and 7/8 yard markings, are not uncommon. The graduations extended completely across the stick, and were usually devoid of embellishment lines. The figures were in the form of proper fractions (1/4, 3/8, etc.), either superimposed on these lines, or placed next to them. Before about 1925 it was customary to graduate these scales right to left; after that date left to right seems to have been the standard.

The yard graduations on the one folding yardstick, the No. 66, differed in two ways from those on the other Stanley yard sticks. On this rule the scale was graduated in both directions, with markings and figures running from both left to right and from right to left, and two equally spaced embellishment lines were added. The right to left graduations were above the upper line and the left to right graduations were below the lower line; the space between them was used for decorative asterisks emphasizing the graduation lines. Both sets of figures used the graduation lines as the separators for their fractions.

Another variation of the fractions of a yard scale was to graduate the first quarter of the scale in inches, and then switch to 8ths of yards for the remainder of the three feet. This is the scheme which was used on the Nos. 450 and 550 counter measures.

32nds of Inches

Although it was not Stanley policy to graduate the wood parts of their rules with markings any finer than 16ths of inches, this was not a limitation when graduating brass caliper and extension slides. Some rules had 32nds graduations standard on their slides, others had them as an option (see Special Order Standard Rules, Chapter 4): ultimately they became standard on all slides.

The system of accenting markings in this scale was a simple extension of that used for 8ths and 16ths graduations, with the 32nd markings slightly shorter than the 16ths.

Metric Graduations

This is the well-known system of measurement which was invented by the French Academy shortly after the revolution of 1789–1799, and spread across Europe by the conquests of Napoleon.

Based on a fundamental unit called the meter (=39.37 inches), subdivided into 100ths (centimeters) or 1000ths (millimeters), it had become the only legal standard in most European countries by the mid-1800s, was in general use in their various colonies by about 1900, and today is almost universally used outside the English-speaking world.

Stanley rules with metric graduations were marked every millimeter, with every 5th line emphasized by making it slightly longer. Every 10th line was made even longer still, and was marked with figures representing the distance from the beginning of the scale in centimeters (11, 12, 13, etc.).

It was customary in Europe to mark scales left to right, instead of right to left as in the United States. At various times Stanley marked their metric scales both ways. Rules which appeared in the catalogue seem to adhere to the European system; custom metric rules, on the other hand, are usually figured right to left, like the ordinary foot/inch rules.

Other 1 : 1 Scales

Several of the 1:1 scales were in units other than inches/fractions or centimeters/millimeters. These were usually special purpose scales used in a particular trade, and were divided in the traditional units of that trade.

Printers' Measure

"Printers' Measure" is not a single scale, but rather the set of scales, one per size of type, which a printer uses in laying out pages and composing. It is much easier for the printer to do all his measuring in units of type size rather than in abstract units like inches or millimeters.

Type is sized in units called "points," where one point is equal to .013837 inches (approximately 1/72 of an inch). In the nineteenth century, each commonly used size had its own name, and its scale would be labeled with that name. The names and sizes of the most frequently used types are listed in Figure 25 below:

Name of Type Size	Type Height in Points	Approximate Height in Inches
Pearl	5	7/100 Inch
Ruby	5½	5/64 Inch
Nonpareil	6	1/12 Inch
Minion	7	1/10 Inch
Brevier	8	1/9 Inch
Bourgeois	9	1/8 Inch
Long Primer	10	9/64 Inch
Small Pica	11	5/32 Inch
Pica	12	1/6 Inch
English	14	1/5 Inch
Great Primer	18	1/4 Inch

Figure 25: Type Names and Sizes

Another scale that was frequently encountered in typesetting and composition is Agate. An Agate is 1/14 of a "column" inch (slightly more than 5 points), and is the unit used by printers to measure the depth of non-text copy such as advertising and illustrations. A two column advertisement, 4 inches deep, for instance, would be described as being "2 columns by 56 Agate lines."

The usual method of graduating printers' scales was to make a mark for each line of that size type (e.g., the marks on the Pica scale would be 12 points apart, those on the Minion scale 7 points, etc.). Every fifth graduation line would be accented by extending it past the single embellishment line, and would be figured with the number of lines it indicated (5, 10, 15, etc.) The intermediate lines would all end at the embellishment line.

Watch Glass Size

Scales of watch glass size were established by watch manufacturers to classify the diameters of the front crystals for the different sizes of watches which they made. They would be used by a watch repairman to measure an existing crystal when selecting a replacement, or when ordering a new one from the factory.

It is not known which of these scales Stanley used on the No. 214 watch glass gauge in 1862. There were a great many different scales, with little, if any, standardization among them. Usually they were calibrated not in fractions of an inch or millimeters, but in the standard crystal sizes (#3, #4, #5, etc.) of a particular maker. Since every maker had a different set of standard crystal sizes, every scale would be different.

Fortieths of an Inch (Button Size)

Fortieths of an inch were the traditional units used to measure and size buttons and buttonholes when sewing and tailoring, and Stanley accordingly placed a 40ths scale on the caliper slide of its boxwood and ivory button gauges for this purpose.

This scale had all graduations the same length, except for every 10th line (coinciding with the 1/4 inch points), which was made extra long. These accented lines were figured with the number of 40ths from the beginning of the scale (30, 40, 50, etc.). This scale was never augmented with embellishment lines.

Linear Scales, Not 1 : 1

A second group of scales graduated on Stanley rules were the linear scales which were graduated in units other than full scale. This group included scales such as the patternmakers' shrinkage scales, where all the "inches" were slightly oversize, to compensate for the shrinkage of the casting during cooling, and drafting scales, which were graduated into scale feet and inches to compensate for the reduced size of the drawing as compared to the size of the actual object.

These were really types of computing scales, each including in its graduation interval some factor which represented the difference between its nominal spacing and the size of the object which it represented. Without these scales, it would have been necessary to take a dimension in true inches, and then apply the needed factor arithmetically, a much slower and more error-prone process.

Drafting Scales

Drafting scales are special scales used to make measurements on scale drawings, drawings where the dimensions of the drawing are some fixed percentage of the dimensions of the object being depicted.

An example of this would be a drawing with a scale factor of 1/2"=1', that is, where each 1/2 inch on the drawing represents 1 foot on the subject of the drawing. Scale drawings are always used in architectural work, and are common in mechanical design and engineering work, where the object drawn is either too large or too small to be conveniently represented by a 1:1 (full scale) drawing.

A typical drafting scale will be divided into intervals equal to scale feet (1/2 inch intervals in the case of the scale factor mentioned above), and then have one or more of those "feet" subdivided into scale "inches" (1/24 inch intervals in this case), with the lines representing 3, 6, and 9 "inches" made longer to emphasize them.

Whether building furniture or houses, carpenters would frequently have to deal with scale drawings, and drafting scales in common scale factors were frequently marked on rules for their convenience. The exact details of the drafting scales on any particular rule would vary as a function of factors like the rule patterns, its size, or the preference of the designer.

Two foot, 2 fold rules, for instance, had four drafting scales, each 9 inches long, with scale factors of 1/4, 1/2, 3/4, and 1 inch per foot. On rules with slides these scales were placed on the same side of the rule as the slide, on the other leg, and were graduated from left to right, beginning near the main joint. On rules without slides they were placed so that two were on each leg, adjacent to the inner edge, and were graduated from right to left, beginning near the tips. The most common scheme of graduation was to divide the first 3 inches of each scale into scale feet and inches, and the remaining 6 inches into scale feet only. The only exception from this scheme was the No. 4; for some reason, on this rule only the first scale foot of each scale was divided into inches, and the remainder, 8 or more inches, into scale feet only (as is the custom in graduating drafting scales today).

The 2 foot, 4 fold rules also had four scales with the same scale factors, but they were shorter (4 inches), and arranged slightly differently. On these rules the drafting scales were located on the inside surface of the rule, one scale per stick, adjacent to the inside edge. These scales were graduated from left to right, and, as on the No. 4, only one scale foot (the inch per foot scale, where it was divided into 6 intervals representing 2 inches each). Also, in accordance with contemporary practice the 0 point of each scale was not at the end, but 1 scale foot to the right of it, at the right hand end to the right of this zero point, the inches to the left, thus making a little easier the process of converting the ordinary form of scale dimension (X feet, Y inches) into a measurement.

The two beveled edge 2 foot, 4 fold rules (the Nos. 53½ and 86½) were graduated identically to the ordinary 4 fold rules, except for the scale factors of the four scales. On the bevel edge rules, the scale factors of the four drafting scales were 1/8, 1/4, 3/8, and 1/2 inch per foot, instead of 1/4, 1/2, 3/4, and 1 inch per foot. On each scale the leftmost foot was divided into 12 scale inches (except for the 1/8 inch per foot scale, where it was divided in 6 scale intervals of 2 inches each).

There were only two drafting scales on the No. 036 Combination Rule, for scale factors of 1/4 inch and 1/2 inch per foot, each four inches long. On these scales 0 was at the right hand end, and the foot between 0 and 1 subdivided into scale inches (on the 1/2 inch scale) or scale 2 inches (on the 1/4 inch scale).

The Nos. 172 and 174 drafting scales offered between 1862 and 1865 had a large number of scales for different scale factors, but their arrangement and details are unknown. These rules were never illustrated in any catalogue, and no identifiable example of either has ever been found.

Shrinkage Graduations

Shrinkage graduations are a special form of inch scale used by patternmakers in making the patterns for foundry use. A shrinkage scale has all the appearance of an ordinary inch scale, but differs in that it is between 1/4 percent and 1½ percent oversize. By using this oversize scale the patternmaker can compensate for certain shrinkages (hence the name) which occur during the foundry process.

Shrinkage scales were graduated in 8ths, 10ths, 12ths, and 16ths, and at different times, on different shrinkage rules, were sometimes accompanied by embellishment lines. Accenting of major divisions by lengthening them was exactly as on ordinary inch scales.

Additional information on shrinkage graduations and shrinkage rules can be found in the section SHRINKAGE RULES, Chapter 5.

Octagonal (4 Square) Scales

This is a pair of scales, the E (Edge) and M (Middle) scales, designed to allow a carpenter to determine how much to cut off each corner of a square to make it an octagon. The Rabone instructions for the carpenters' sliding rule (Ref. 111) describe these scales as being used by mast and spar makers, but they were probably more often employed to solve problems in brace length when framing buildings.

These scales were always marked on 2 foot, 2 fold rules, and, while both were slightly more than 7 inches long, had different graduations. The E scale was graduated from 0 to 24; the M from 0 to 32. Both were used to find the point at which the bevels must begin in order to turn a square into an octagon, but used different points on the square as a reference. The E scale located that point with reference to the edge (corner) of the square; the M scale with reference to the middle (the center point of face). The length of the side of the square would be found on either the E or the M scale, and opposite that point on the ordinary inch scale would be the distance from the desired reference point to the point at which the bevel must begin.

These scales were always placed one on each leg on one side of the rule, and were both graduated from the end of the leg toward the joint. They were not placed at the edge of the leg, but on the surface between the two ordinary scales, and used one of their two embellishment lines to define their apparent edge. On these scales the minor graduation, which only extended to the second embellishment line, represented 1/2 inch; the inch markings extended slightly beyond the line, and the 2 inch markings all the way past the figures with which they were marked.

Board Scales

Board scales were computing scales, calculated to convert the length and width of a 1 inch board into its equivalent area in board feet (a board foot is 1 square foot of wood 1 inch thick). Each scale would be calculated and graduated for a specific length board (12 feet, 14 feet, etc.), such that opposite the width of any board of that length would be a figure representing its equivalent in board feet. The scale for 14 foot boards, for instance, would have its 7 board feet marking 6 inches from the beginning of the scale (6 inches divided by 12, times 14 feet, equals 7 board feet), its 8 board feet marking at 6.86 inches, etc.

These scales would be marked on a board stick, from 1 to 4 to each surface, depending on the stick's cross sectional shape. Flat sticks would have 3 on each face (for a total of 6); square sticks would have 4 per face (fo a total of 16); octagonal would have either 1 or 2, depending on the thickness of the stick (giving either 8 or 16). When multiple scales were marked on a single face, they would be separated by embellishment lines. Board scales were graduated at each whole board foot only (1, 2, 3, etc.), with a line running from edge or embellishment line to edge or embellishment line, and always ran from left to right, beginning at the head or hook of the stick.

Additional information on board scales and board sticks can be found in the section BOARD MEASURE AND BOARD STICKS, Chapter 5.

Nonlinear Scales

The most complex scales put on Stanley rules were the nonlinear scales, scales where the interval between graduations of equal difference are not uniform, but vary along the length of the scale. These were computing scales of an advanced type, involving functions more complex than simple multiplication by a constant, etc.

Some of these scales, like the Doyle's log scale and those on the Gunter's slide, were based on mathematical formulas. Others, like Scribner's log scale and the scales on the gauging rods, were derived from measured data, and did not represent any particular mathematical expression.

These scales were almost invariably cut by hand, using steel patterns in the manner described in Chapter 2. It is possible that methods to do this by machine were developed after the turn of the century; machines capable of cutting nonlinear scales, based on the principle of the pantograph, with the ganged graduating knives controlled by a stylus guided across a template, were well known by that time (Ref. 118). No evidence has been found to either support or refute this possibility, however.

Log Scales

A log scale is a specialized scale designed to simplify the problem of estimating how much sawn lumber can be obtained from a given log. Sticks calibrated with these scales are employed by lumbermen and sawyers for selecting trees for cutting and as a basis for payment at the mill, and are widely used, even today.

The basis for a log scale is a log rule, a table or formula which has been developed to relate log length and diameter to the number of board feet of lumber which can be sawn from such a log. A large number of such rules have been published over the years, to deal with different types of trees and varying sawmill methods (Ref. 131). The two rules which are known to have been marked on Stanley log canes are Doyle's (1825) and Scribner's (1846).

Log scales can be readily distinguished from board scales by a cursory glance. The graduations on log scales are uniformly spaced 1 inch apart, with each graduation line figured with the nearest integral number of board feet which it represents. The graduations on board scales are also linearly spaced, but with a different spacing for each scale (1 inch apart for the 12 foot scale, approximately 1.091 inch apart for the 11 foot scale, etc.), and are marked such that the figures on adjacent lines differ only by one.

A more complete description of log rules can be found in the section LOG RULES AND LOG MEASURES, Chapter 5.

Gauging and Wantage Scales

Gauging and wantage scales are specialized computing scales designed to simplify the work of the gauger, the individual responsible for measuring the capacity of liquid containers, and the quantity of liquid therein, if less than full.

The gauging scale is used for capacity measurement. It is designed to convert the linear distance from the filling hole in the side of a barrel to the corner opposite, and to read the capacity of the barrel directly in gallons. It is a simplistic form of measurement, depending heavily for its accuracy on the fact that barrels tend to have proportional dimensions whatever their size (that is, the head diameter is always the same fraction of the barrel length, etc.). A gauging scale less than 3 feet long could measure the capacity of barrels up to 120 gallons, one about 3½ feet long was sufficient for barrels up to 180 gallons.

Measuring rods graduated with gauging scales were useful for quick measurements where great precision was not required, and were widely used for many years.

The wantage scale was another computing scale, used to convert the distance from the liquid surface in a barrel to the filling hole into a figure representing the amount of liquid that the barrel "wants" of being full (hence the name) in gallons. These scales were barrel-specific, each size barrel requiring a different scale, and usually would be marked on wantage rods in groups, like the multiple scales on board sticks.

For more information on gauging and wantage scales, see GAUGING AND WANTAGE RODS, Chapter 5.

Degree and Pitch Scales

The No. 036 combination rule was

marked on its blade with scales for both degrees and pitch. Both are measures of angular distances, but are expressed in different terms. Degrees is the familiar system of angular measurement we all learn in school; pitch is the framers' and millwrights' expression for slope, in which the angle relative to the horizontal is expressed as the ratio of rise (change in height in a given horizontal distance) to run (the horizontal distance). Workers in these trades did much of their angular measurement using the two legs of their framing square (Ref. 125), and it was easier and more natural for them to think in terms of rise and run, and of pitch as the ratio of the two. Thus where a mathematician would refer to a slope of 30 degrees, a house carpenter would speak of a pitch of nearly 7 inches per foot.

The tongue of the combination rule was marked with pitch and degree scales, one on each side. Although these are linear functions, the geometry of the rule required that these scales be nonlinear, being compressed as the measured angle increased. The degree scale was marked at each degree, with every 5th degree accented by extending its graduation line past the single embellishment line. The 10 degree lines were made longer still, and every 5th degree was marked with a figure (5, 10, 15, etc.).

The pitch scale was graduated in 8ths of an inch in inches of rise, with only the half and whole inch lines extending past the embellishment line, and only the latter marked with figures. Since the length of each leg of the No. 036 (the run) was 6 inches, this scale could be read directly in inches per 1/2 foot. It was probably more common, however, for the user to mentally double that number so that it was expressed in inches per foot, the more common way (along with feet per foot) of stating pitch.

The No. 036 had a second degree scale marked on the inside edge of the leg containing the spirit level. This scale was designed for use in setting the angle of the tongue relative to a line normal to the leg containing it, as in when the tongue functioned as a protractor or bevel gauge. This scale was very simple, consisting of a series of short (1/8 inch) transverse lines in the middle of the edge surface, devoid of embellishment lines, and with degree figures adjacent to every fifth mark. The location of this scale, and the geometry of its use was such that it could be linearly graduated, but still indicate angular degrees correctly.

The "Gunter's" Slide

"Gunter's Slide" was the name applied during the eighteenth and nineteenth centuries to the logarithmic slide rule, after Edmund Gunter (1581–1626), the English mathematician. Gunter, who had a knack for finding practical applications for the mathematical discoveries of his day (he was the inventor of the Gunter's Chain, used in surveying, and two navigating instruments, the Gunter's Quadrant, for taking observations, and the Gunter's Scale, for performing computations), was popularly considered to be the inventor of the slide rule, although his actual contribution was only the invention of the logarithmic scale. The true inventor, the man who conceived of two or more logarithmic scales which could be shifted (slid) relative to one another, was William Oughtred (1574–1660), another mathematician.

The specific form in which the slide rule was applied to folding rules was developed about 1677 by a third Englishman, Henry Coggeshall. Coggeshall, who was particularly concerned with timber mensuration, arranged the scales on his slide rule to suit that class of problem, and it was this arrangement, or a minor variation of it, that was generally used on carpenters' and engineers' (sliding) rules.

Coggeshall's pattern had four scales: the A scale, on the body of the rule above the slide, the B and C scales, on the upper and lower edges of the slide, and the D scale, on the rule body below the slide. The A, B, and C scales were what today is referred to as two-cycle scales, non-folded. That is, each scale began at 1 on the left, progressed through 2, 3, ..., 8, 9, to 1/10 in the middle, and then continued, repeating the sequence again, to 10 on the right. Two copies of this scale (the B and C scales) were required on the slide to compensate for the lack of a cursor (the sliding vertical line which is a feature of all modern slide rules). The D scale was a single-cycle scale, but was marked differently on the two different types of rules; on the carpenters' rule it was "folded" at the value 4, while on the engineers' rule it was non-folded, beginning and ending at 1.

Additional information on the two forms of Gunter's slide will be found in tHE CARPENTERS' (SLIDING) RULE and THE ENGINEERS' (SLIDING) RULE, Chapter 5.

Tables

In some cases data was put on rules in the form of tables, with the vertical and horizontal lines scribed, and different values stamped into the resulting boxes, one figure at a time. The most elaborate of these were the tables of gauge points marked on the Nos. 6 and 16 engineers' (sliding) rules, but there were other tables for such purposes as board measure calculations and hat size conversion as well.

Data Tables were expensive to apply to rules, involving as they did a great deal of slow, one figure at a time hand work. The last rules with data tables were the 4 fold rules with board measure tables, all of which were discontinued in 1917 and 1919.

Weight and Measures Tables

These were the "Tables for ascertaining the Weights of all sizes of Iron, Steel, Copper, Brass, Lead, &c." which were marked on the No. 58½ 6 fold rule from 1860 through 1862. This rule was never illustrated in any catalogue, and no identifiable example is known to exist, so the only information we have about these tables are is description above, taken from the catalogue. Probably they were similar to the weights and measures tables marked on some English rules during the nineteenth century (Refs. 16 and 19), but until an example is found, this is only speculation.

Board Measure Tables

A board measure table is a table of equivalent board foot values for 1 inch thick boards of various lengths and widths. Each column in such a table would represent a different board length (3 feet, 4 feet, etc.), and each row a different width (7 inches, 8 inches, etc.), and the contents of any board in board feet and board "inches" (12ths of a board foot) could be found in the space in the table corresponding to its length/width.

Stanley actually marked two tables on its rules: a small table, for boards 1 to 12 feet long and 6 to 12 inches wide, and a large table, for boards 1 to 19 feet long and 13 to 19 inches wide. These tables were always placed side by side, with the small table on the left. When marked on folding rules they were half on each leg (on the inside surface if it was a 4 fold rule), and thus could only be used when the main joint was closed.

The tables were too large vertically to be marked on a 1 inch rule, and were only offered on rules at least 1$^{5}/_{16}$ inch wide: a bench rule, two 2 fold rules, and four broad 4 fold rules.

Additional information on board measure tables and the rules on which they were marked can be found in the section BOARD MEASURE AND BOARD STICKS, Chapter 5.

Hat Size Table

The hat size table was a small three column table stamped on the back of the No. 212 hatters' rule. The three columns of this table were labeled SIZE (the front to back inside diameter of the hat), HEAD (the circumference of the customer's head), and HAT (the circumference of the hat outside the ribbon). This table allowed the clerk to measure either the customer's head or his old hat to determine what size new hat to try. The diameter would be measured with the slide of the rule, the circumference with a measuring tape.

More information about this table, and about the hatters' rule, can be found in the section HATTERS' RULES, Chapter 5.

Engineers' Tables

These were the elaborate tables of gauge points devised around 1811 by Josiah Routledge for his so called "engineers' sliding rule." The most popular of a number of similar "engineers'" rules introduced during the early nineteenth century (Ref. 143), Routledge's rule consisted of a 2 foot, 2 fold rule marked with his tables and equipped with a Gunter's slide (a simple form of slide rule). Consisting of 168 data values in five categories, the tables provided the core of the material needed to solve many of the primitive engineering problems of the time (Ref. 146). Many rule makers, including Stanley, manufactured rules of Routledge's pattern until after the turn of the century.

These tables are probably the most elaborate markings made on any Stanley rule. Occupying a space only ¾ inch wide by 10½ inches long, they were framed by a grid of vertical and horizontal embellishment lines, and stamped totally by hand, one digit at a time. It is probable that only the most skilled and painstaking workers were used for these rules, and that at best each worker could only produce a few each day.

It has been found (Ref. 147) that these tables were notoriously prone to error, with an average of probably one misstamped figure on every rule. The apparent cause for this is a combination of worker error (the figures are very small, and the slightest inattention can cause mistakes) and the tendancy of rule manufacturers (including Stanley) to copy such tables off another maker's rule, mistakes and all.

Additional information on Routledge's rule and the engineers' tables can be found in the section THE ENGINEERS' (SLIDING) RULE, Chapter 5.

CHAPTER 4 Custom Rules and Rule Accessories

Custom Rules

The more than 230 different rules described in Chapter 5 are those which were "standard" products, rules which were listed in one or more of the Stanley tool catalogues between 1855 and 1975. Of almost equal interest, although nearly impossible to catalogue, are the many more nonstandard rules which were available only on custom order, or were for use in other fields and were only advertised elsewhere. Some of these custom rules were almost identical to standard rules, differing only in some minor detail. Others were totally unique, unlike any rule in the standard product line.

It is beyond the scope of this book to catalogue all known and possible Stanley custom rules (it would require a good sized book to illustrate just the more than 500 Special Order Standard Rules alone). It is possible, however, to classify them based on their relationship to the Stanley line of standard rules, and to illustrate a few examples from each category and comment on some of their features.

Generally speaking, custom rules fell into one of four groups: special order standard rules, advertising rules, full custom rules, and purchaser modified rules.

Special Order Standard Rules

These were rules which were relatively minor variations of standard rules. While not individually listed in the catalogue, they were mentioned in footnotes as being available on special order (sometimes at slight extra cost). Some were only minor differences from standard graduations; others were construction variations, or rules designed for use in other areas of the world. The features which made the rules in this category special were:

Board Measure Tables (1860–1867)

These were the same tables which were marked on the Nos. 22, 35, 79, 80, and 82 rules, and could be ordered marked on "any" rule for an extra charge of $2.00 per dozen. It is unlikely that these tables were actually available on standard width (1 inch) and narrow (3/4 inch) rules; their size was such that they would not fit on any rule less than 1¼ inch wide.

English Markings (1909–1932)

"English" markings was the name for the scale arrangement where the figures run from left to right (that is, with the scale beginning on the left, when the rule is held so that the figures are viewed in the upright position), instead of the almost universal right to left arrangement of American rules. Left to right graduations was the customary arrangement on English (hence the name) and Continental rules, and were sometimes preferred by craftsmen who had been trained there.

It is not known which rules could be had with this feature; the catalogues only state that such rules were "available." Since this was a fairly simple modification, involving only the hand stamped figures applied after graduating, it was probable that any rule could be ordered this way.

Beginning in 1926, rules with English markings were given their own distinctive number, consisting of the basic rule number followed by the suffix E (e.g., No. 62E) or, if also marked with metric graduations, EM (e.g., No. 62EM).

Figure 26: No. 69E, 1926–1932 *(Charles and Walter Jacob Collection)*

Figure 27: No. 98E, 1926–1932 *(Charles and Walter Jacob Collection)*

Metric Graduations (1867–1932)

This feature consisted of graduating one side of the rule (the outside, inthe case of 4 fold rules) in centimeters and millimeters instead of inches and fractions and from left to right, in accordance with common European usage. Until 1874 this option was described as "French" graduations, the term "metric" only coming into use after 1877, as more and more countries adopted this system as standard.

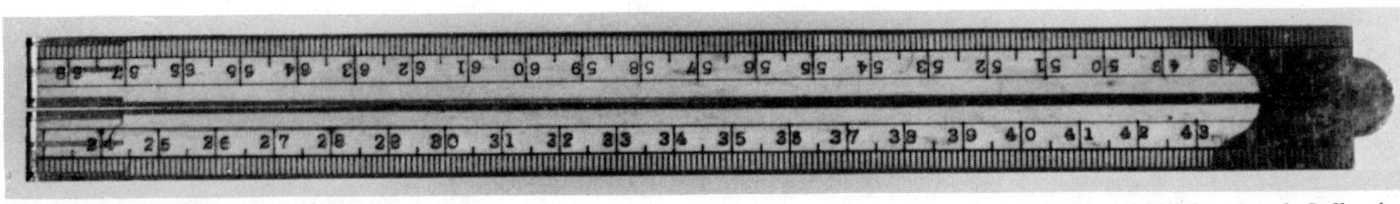

Figure 28: No. 66¾M, 1912–1925 *(Charles and Walter Jacob Collection)*

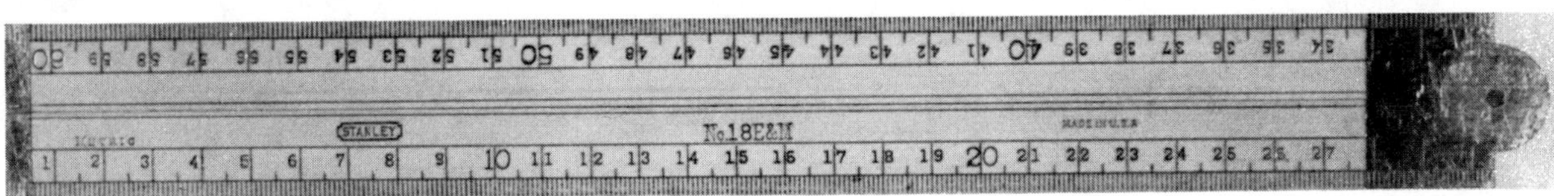

Figure 29: No. 18E&M, 1926–1932 *(Charles and Walter Jacob Collection)*

It is not clear whether all rules were available with this option. Until 1905, it was simply stated in the catalogue that rules with these graduations were "available." In 1909, this was changed to read "any rule ...," but shrinkage rules, extension sticks, and a few other miscellaneous rules were excluded. Then in 1912, the offer reverted back to its original scope, simply that such rules were "available."

Beginning in 1926, metric graduations were linked to "English" markings (see above); from that time on rules could be ordered with either the suffix E (English markings on both sides) or EM (English markings on one side, and metric graduations on the other).

After 1932, the availability of custom metric graduations was no longer mentioned in the catalogue.

Spanish Graduations (1865–1902)

Spain had legally adopted the metric system in 1859, and the Spanish-speaking nations of South America and Central America had followed that country's lead within a few years thereafter. Prior to that time they had used a system based on a foot (in Spanish, "pie") equal to 11.13 English inches, divided into 12 "pulgadas" (literally "thumbs"), each subdivided into 12 "lineas" (lines).

This old Spanish system persisted in use long after it had been officially replaced, however, especially in the Americas, and it was this market which Stanley was addressing with this offer.

By about 1900, however, the old system had finally yielded to the new (with some encouragement from the countries involved; Mexico, for instance, passed a law in 1896 which made the use of the metric system "obligatory"), and in 1905 this option was no longer offered.

Nonstandard Shrinkage Allowances (1898)

The standard length for the Nos. 30 and 31 shrinkage rules was 24¼ inches (corresponding to a shrinkage allowance of 1/8 inch per foot). In 1898, however, a note in the catalogue stated that they could also be furnished with a shrinkage allowance of 1/16, 1/10, 3/32, 3/16, or 1/14 inch per foot, on special order.

Two years later, in 1900, the Nos. 30½ and 31½ shrinkage rules were introduced, available as standard products in all of these allowances plus others. This eliminated the need for special shrinkage allowances on the Nos. 30 and 31, and consequently this option was no longer offered.

Optional Shrinkage Graduations (1900–1940)

The Nos. 30½ and 31½ shrinkage rules came standard graduated to scale in 8ths, 10ths, 12ths, and 16ths of inches, but from the time of their introduction until 1929 were also available graduated in only 8ths and 16ths at no extra cost. This must have been a fairly popular option; about one in eight of these shrinkage rules examined by the author have been found to be graduated in this fashion.

Right Hand and Left Hand Calipers (1905–1940)

Up until 1932 the standard configuration for caliper rules was to have the caliper left hand (see SLIDES AND CALIPERS, Chapter 2). From 1905 on, however, it was stated in the catalogue that for a small fee (varying at different times from 25 to 60 cents per dozen) these rules could be ordered with a right hand caliper instead.

In 1934, Stanley did an about face, and adopted the right hand caliper as the standard configuration. At the same time, the number of rules available in the optional configuration (now a left hand caliper) was reduced to one, the No. 36½, which could be ordered in the left hand form as the No. 36½ (Special). It is not known whether there was any charge for this variation.

In 1941, the No. 36½ (Special) was made a standard product as the No. 36½L, the No. 36½ being renumbered as the No. 36½R at the same time. From that time on, special order caliper rules were no longer available.

32nds Caliper Graduations (1905–1922)

Up to 1922, the standard graduations on the brass slides of almost all caliper rules was 16ths of inches (the only exceptions were the Nos. 76C and 83C, which came standard with 32nds). Beginning in 1905, however, any other caliper rule could be ordered with the slide graduated in 32nds without additional charge. This option was available as long as calipers were graduated in 16ths of inches; when 32nds of inches became standard in 1925, it was obviously no longer relevant, and was dropped.

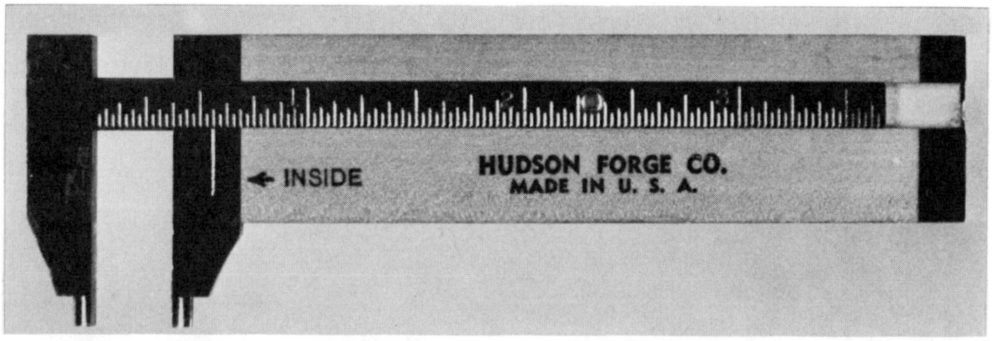

Figure 30: Custom No. 136 Made for the Hudson Forge Co. *(Author's Collection)*

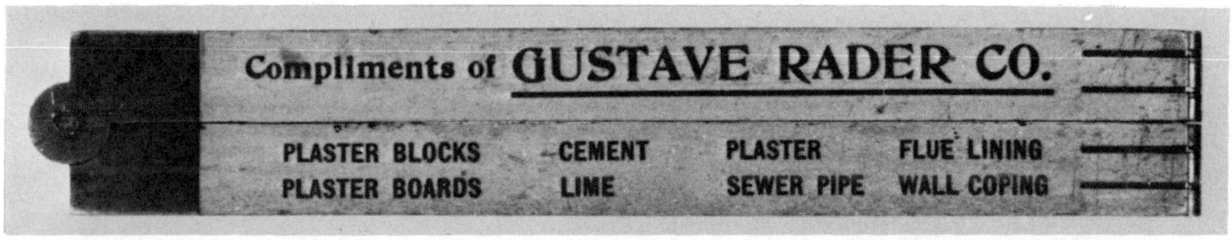

Figure 31: Custom No. 61 Made for the Gustave Rader Co. *(Author's Collection)*

Figure 32: Custom No. 36 Made for the Reed & Prince Mfg. Co. *(William Baader Collection)*

Advertising Rules

These were rules manufactured by Stanley in limited quantities for some other company for advertising purposes. The Stanley trademark, and possibly one or more scales, would be omitted, and the purchaser's trademark and message would replace them. The purchaser would then sell or give away these rules to advertise their business, or to promote good will.

Three examples of this class of custom rule are shown here. Two, the custom No. 36 and the custom No. 136, replaced only the Stanley trademark and rule number with their own name. The third, the custom No. 61, replaced all of the markings on one side of the rule with their own.

Full Custom Rules

These were rules which were markedly different from standard products. Some, while retaining the shape of a standard rule, were made of nonstandard materials, or had different graduations. Others, more extreme variations, had different lengths or shapes, or incorporated unique features. The one feature all had in common was that they were never mentioned in any Stanley woodworking tool catalogue, even in a note mentioning their "availability." Many, such as the Milk Can Dipstick (see below), were not intended for use by woodworkers, and may have only been advertised elsewhere, in media appropriate to their intended use.

Custom Metric Length Rules

Stanley custom rules in metric lengths, while not common, are sometimes encountered. Two examples are shown here: a 40 cm., 4 fold common rule, and a 20 cm., 2 fold caliper rule. Neither of these rules corresponds with any metric length rule ever listed in any Stanley catalogue, nor is either marked with any rule number. Both, however, are marked with the Stanley Rule & Level Company trademark, and would appear to be of pre-1900 manufacture.

In contrast to any other pre-1934 Stanley rule bearing metric graduations, these custom metric length rules are graduated only in metric (on both sides). In all other cases (the metric length and English length rules of 1877—1879, the 1 foot school rule of 1919—1932, the meter sticks of 1926—1941 and of 1926—1943, and the special order rules with metric graduations of 1867—1932) the metric graduations were only on one side, the other side being divided into inches and fractions. No information is available as to why these rules should be different; perhaps they were intended for sale in Europe or South America, where dual graduations were not necessary.

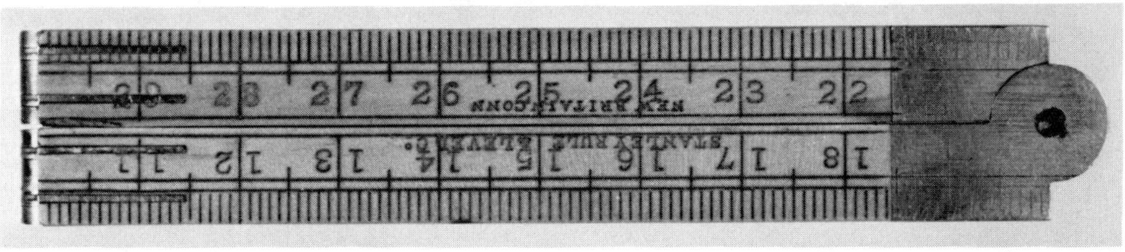

Figure 33: Stanley Rule, 40 Cm., 4 Fold

(Author's Collection)

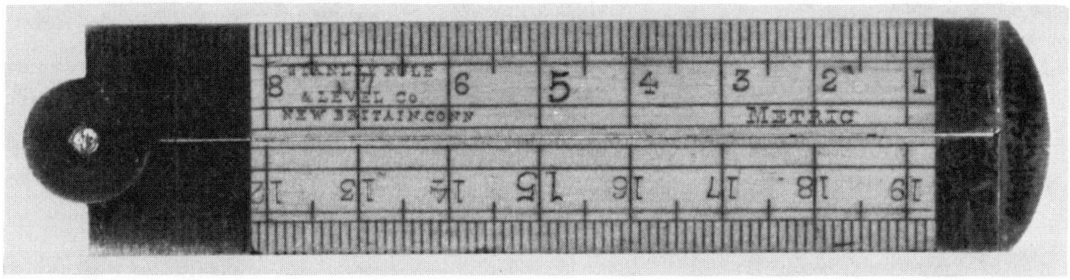

Figure 34: Stanley Caliper Rule, 20 Cm., 2 Fold

(Charles and Walter Jacob Collection)

Custom Graduations on Standard Shape Rules

It was relatively inexpensive to make a custom rule when its shape and dimensions were the same as or nearly the same as those of some standard product. In such cases, the only special handling that such rules might require would only be in the final, finishing stage of their manufacture. The joint making and framing could be done on the same production line as the standard product, and the cost of the resulting rule be as low as possible.

Three examples of such rules are shown here: a milk can measure, a custom-graduated bevel-edged architects' rule, and an ivory cotton staple gauge.

A milk can measure is a dipstick used to measure the liquid level in the cans used by dairy farmers to ship milk to their local processing plant. It would have been used either by the farmer or the pick-up driver to measure the amount of milk picked up each day for record purposes.

This measure is obviously a derivative of the No. 45 gauging rod, cut down to 30 inches length, and graduated with suitable scales for the various standard milk can sizes (the side visible in the illustration is graduated for a "40 Quart Iron Clad Milk Can"). It is marked with the Sweetheart trademark, and therefore must have been made during 1922–1932.

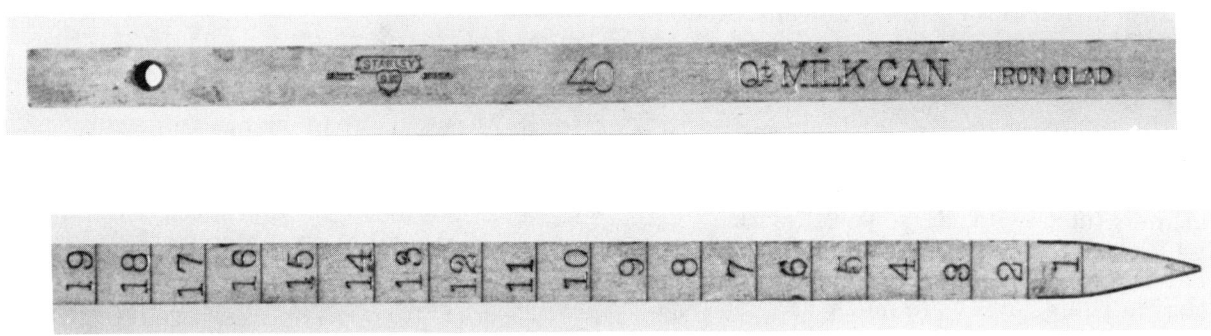

Figure 35: Custom Stanley Milk Can Measure, 1922–1932 *(Roger K. Smith Collection)*

A second custom rule based on a standard product is the bevel edge architects' rule shown in Figure 36. This rule is physically identical with the No. 53½ architects' rule, but has its outside surface graduated in 100ths of a foot instead of the usual inches and fractions.

While not marked with any maker's name, this rule can easily be identified as a Stanley product from the style of the figures and certain construction details. From the shape of the arch joint it can be dated as of post-1909 manufacture, but at the same time it must predate the final cancellation of the No. 53½ in 1943 (Ref. 53).

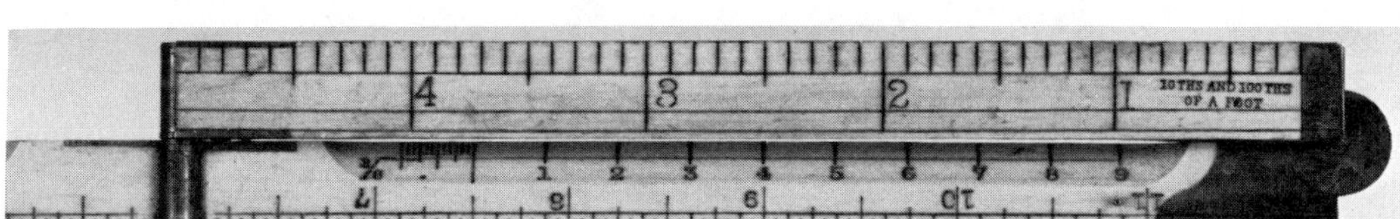

Figure 36: Custom Stanley 4 Fold Architects' Rule *(Author's Collection)*

The third rule in this group, shown in Figure 37, is a standard pattern rule manufactured of a nonstandard material. The No. 299 cotton staple gauge was only offered in the catalogue for one year, 1929, and then only in boxwood. It may have been available informally or on special order for years before that, however, and occasionally been ordered in ivory as a presentation piece, etc.

This rule predates the 1929 listing of the boxwood gauge by at least 8–10 years. This can be seen both by the Stanley Rule & Level Co. trademark, and also by the choice of material; Stanley had ceased the manufacture of ivory rules in about 1918–1920.

Figure 37: Custom Stanley Ivory Cotton Staple Gauge *(Charles and Walter Jacob Collection)*

Custom Rules with New Shapes and Functions

The most extreme forms of custom rules made by Stanley were those rules which were totally different in form from any standard rule. Three of these are shown here: a cordage rule, a slide rule, and a small 2 fold ivory rule.

A Stanley cordage rule, a type of rule never listed in any catalogue, is shown in Figure 38. Cordage rules are special extra wide caliper rules with deep jaws, marked with tables relating the diameter of a cable or hawser to its running weight and breaking strain. These rules were much used by riggers and seamen on sailing ships well into the early twentieth century, for measuring lines and for selecting the appropriate size and type for a given load or application.

Stanley made these custom cordage rules in at least two patterns. The larger, shown here, had a 6 inch by 2½ inch body and caliper jaws 2 inches deep. Its tables gave values for both manila and hemp cordage, for diameters up to 4 5/16 inches. It is marked with the Stanley Rule & Level Co. trademark, but no rule number, on the edge. The smaller size (not shown) had a 4 inch by 2 inch body, caliper jaws 1½ inches deep, and a simple radius in place of the S-shaped curve on the back of the larger rule's caliper jaw.

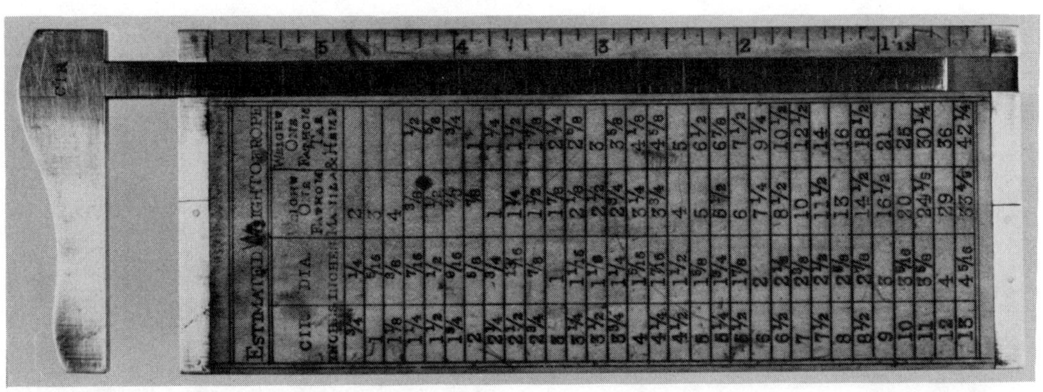

Figure 38: Custom Stanley Cordage Rule *(Charles and Walter Jacob Collection)*

Stanley also made custom rules for other parties. One of these, "Hogg's improved slide rule," is shown in Figures 39a and 39b. This is a slide rule developed for use in the textile industry sometime prior to 1886 by James Hogg of Lawrence, Massachusetts, and sold under his name. It was made by the Stanley Rule & Level Co., however, and the Stanley name is marked directly under Hogg's on the back.

Hogg was a designer or inventor connected with the textile industry (he was to subsequently patent four other inventions relating to spinning and carding machinery). He developed his slide rule by rearranging the four scales found on the Gunter's slide of the engineers' rule, duplicating two of them, and augmenting those six with two other new ones (a linear scale, called the L scale on modern slide rules, and a reciprocal scale, today

Custom Graduations on Standard Shape Rules

It was relatively inexpensive to make a custom rule when its shape and dimensions were the same as or nearly the same as those of some standard product. In such cases, the only special handling that such rules might require would only be in the final, finishing stage of their manufacture. The joint making and framing could be done on the same production line as the standard product, and the cost of the resulting rule be as low as possible.

Three examples of such rules are shown here: a milk can measure, a custom-graduated bevel-edged architects' rule, and an ivory cotton staple gauge.

A milk can measure is a dipstick used to measure the liquid level in the cans used by dairy farmers to ship milk to their local processing plant. It would have been used either by the farmer or the pick-up driver to measure the amount of milk picked up each day for record purposes.

This measure is obviously a derivative of the No. 45 gauging rod, cut down to 30 inches length, and graduated with suitable scales for the various standard milk can sizes (the side visible in the illustration is graduated for a "40 Quart Iron Clad Milk Can"). It is marked with the Sweetheart trademark, and therefore must have been made during 1922–1932.

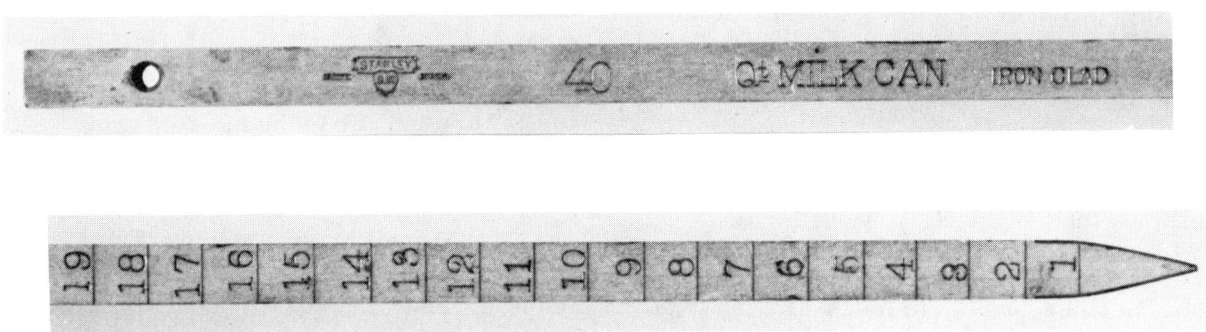

Figure 35: Custom Stanley Milk Can Measure, 1922–1932

(Roger K. Smith Collection)

A second custom rule based on a standard product is the bevel edge architects' rule shown in Figure 36. This rule is physically identical with the No. 53½ architects' rule, but has its outside surface graduated in 100ths of a foot instead of the usual inches and fractions.

While not marked with any maker's name, this rule can easily be identified as a Stanley product from the style of the figures and certain construction details. From the shape of the arch joint it can be dated as of post-1909 manufacture, but at the same time it must predate the final cancellation of the No. 53½ in 1943 (Ref. 53).

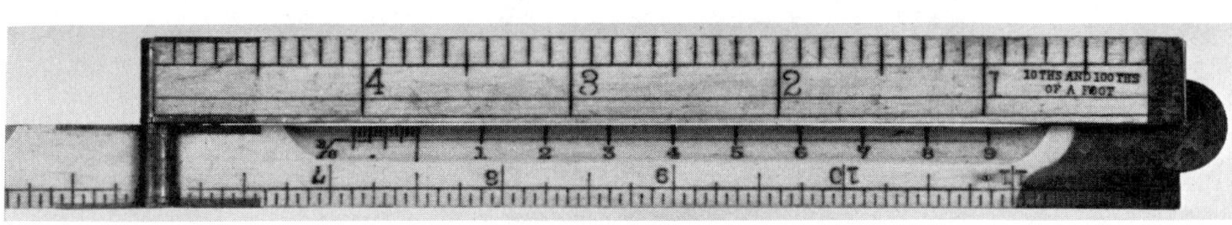

Figure 36: Custom Stanley 4 Fold Architects' Rule

(Author's Collection)

The third rule in this group, shown in Figure 37, is a standard pattern rule manufactured of a nonstandard material. The No. 299 cotton staple gauge was only offered in the catalogue for one year, 1929, and then only in boxwood. It may have been available informally or on special order for years before that, however, and occasionally been ordered in ivory as a presentation piece, etc.

This rule predates the 1929 listing of the boxwood gauge by at least 8–10 years. This can be seen both by the Stanley Rule & Level Co. trademark, and also by the choice of material; Stanley had ceased the manufacture of ivory rules in about 1918–1920.

Figure 37: Custom Stanley Ivory Cotton Staple Gauge *(Charles and Walter Jacob Collection)*

Custom Rules with New Shapes and Functions

The most extreme forms of custom rules made by Stanley were those rules which were totally different in form from any standard rule. Three of these are shown here: a cordage rule, a slide rule, and a small 2 fold ivory rule.

A Stanley cordage rule, a type of rule never listed in any catalogue, is shown in Figure 38. Cordage rules are special extra wide caliper rules with deep jaws, marked with tables relating the diameter of a cable or hawser to its running weight and breaking strain. These rules were much used by riggers and seamen on sailing ships well into the early twentieth century, for measuring lines and for selecting the appropriate size and type for a given load or application.

Stanley made these custom cordage rules in at least two patterns. The larger, shown here, had a 6 inch by 2½ inch body and caliper jaws 2 inches deep. Its tables gave values for both manila and hemp cordage, for diameters up to 4 5/16 inches. It is marked with the Stanley Rule & Level Co. trademark, but no rule number, on the edge. The smaller size (not shown) had a 4 inch by 2 inch body, caliper jaws 1½ inches deep, and a simple radius in place of the S-shaped curve on the back of the larger rule's caliper jaw.

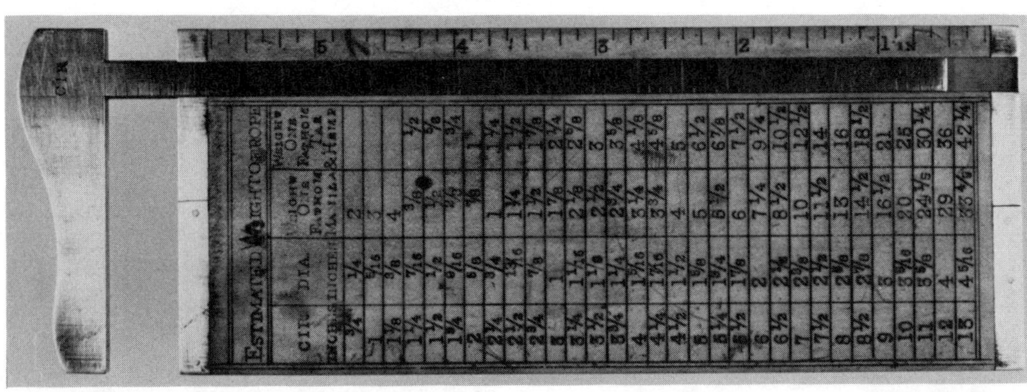

Figure 38: Custom Stanley Cordage Rule *(Charles and Walter Jacob Collection)*

Stanley also made custom rules for other parties. One of these, "Hogg's improved slide rule," is shown in Figures 39a and 39b. This is a slide rule developed for use in the textile industry sometime prior to 1886 by James Hogg of Lawrence, Massachusetts, and sold under his name. It was made by the Stanley Rule & Level Co., however, and the Stanley name is marked directly under Hogg's on the back.

Hogg was a designer or inventor connected with the textile industry (he was to subsequently patent four other inventions relating to spinning and carding machinery). He developed his slide rule by rearranging the four scales found on the Gunter's slide of the engineers' rule, duplicating two of them, and augmenting those six with two other new ones (a linear scale, called the L scale on modern slide rules, and a reciprocal scale, today known as the CI scale).

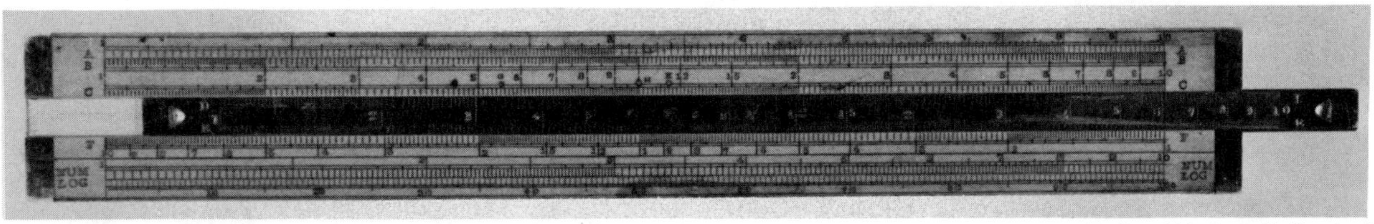

Figure 39a: Hogg's Patent Slide Rule, Front *(Ruben Morrison Collection)*

On the back of his rule Hogg put most of the same tables that were marked on the Nos. 6 and 16 engineers' rules (see THE ENGINEERS' RULE, Chapter 5), omitting only the table of gauge points for pumping engines. Additionally, at three points in the surface of the C scale (at the values 5.5, 10, and 11.4) he inserted tiny brass plugs, similar to those found in the Gunter's scale (see THE GUNTER'S SCALE, Chapter 5), to mark three other gauge points which were frequently used in textile-related calculations.

Hogg also published a book of instructions for use with his rule (Ref. 113) in 1887. This book, 162+ pages long, was divided into two parts. Part I dealt with its use as an ordinary engineers' rule, and was similar in content to books by Chapin (Ref. 103) and others. Part II was devoted entirely to problems and calculations peculiar to textile manufacture (e.g., "To find the Constants used for Gear-Lines on Draft of Spinning-Frames").

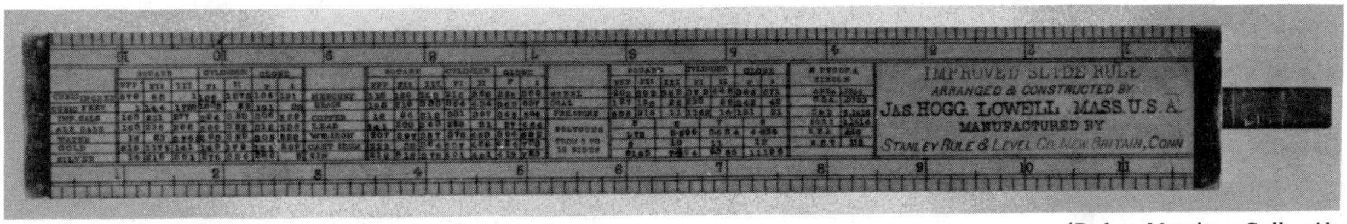

Figure 39b: Hogg's Patent Slide Rule, Back *(Ruben Morrison Collection)*

A third example of extreme custom rule is shown in Figure 40. This is a small brass-trimmed 2 fold ivory rule, marked very clearly with the Stanley name and the rule number (No. 00). No such rule was ever offered in any Stanley catalogue under this, or any other number, but several examples are known to exist, and a significant number must have been made.

It has been suggested that these No. 00 rules were assembled from leftover ivory blanks and joints after ivory rulemaking was discontinued around 1918–1920, and that they were "giveaway pieces," to be given to customers by salesmen to promote good will. No concrete evidence has been found to support this theory, however, and it must remain only a supposition.

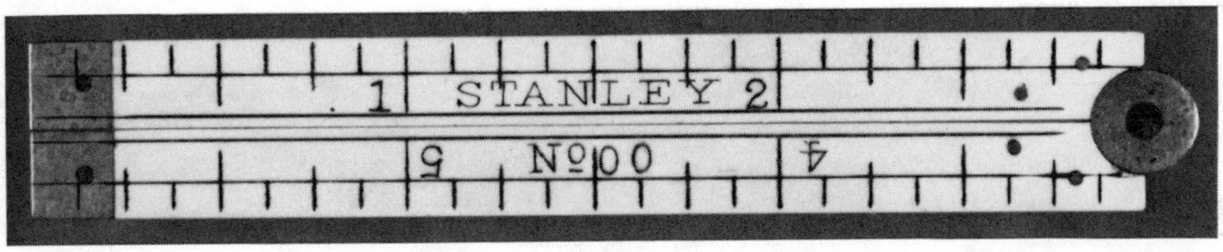

Figure 40: Stanley No. 00 6 Inch, 2 Fold Ivory Rule *(Charles and Walter Jacob Collection)*

Purchaser Modified Rules

It was sometimes the case that rules would be purchased from Stanley as standard products, and then modified by the purchaser and resold under his own name. While not, properly speaking, Stanley custom rules, they are Stanley rules, and are included here so that they can be recognized when encountered.

Nicol's Patent Rule Dividers

This is an example of a rule adapted to also function as a pair of dividers. A Stanley 2 foot, 4 fold rule has been equipped, near the middle joints, with a pair of retractable divider points, and has had the main joint pivot pin replaced with a thumbscrew, so that it can be locked.

This adaptation is the invention of R. Nicol of Rockford, Ill., and was patented by him on June 25, 1872 (U.S. Patent No. 128,417, entitled "Improvement in Combined Rule & Dividers").

It is not known how extensively these dividers were marketed. While the rule shown here is the only known example of this invention, it differs from the illustration in the patent in several details (in that drawing the divider points have a screw retracting mechanism, and the invention is depicted installed on a 2 foot, 2 fold rule), and is almost certainly not the patent model. This would indicate that it actually was put into production, but how many were made, and for how long, is not known.

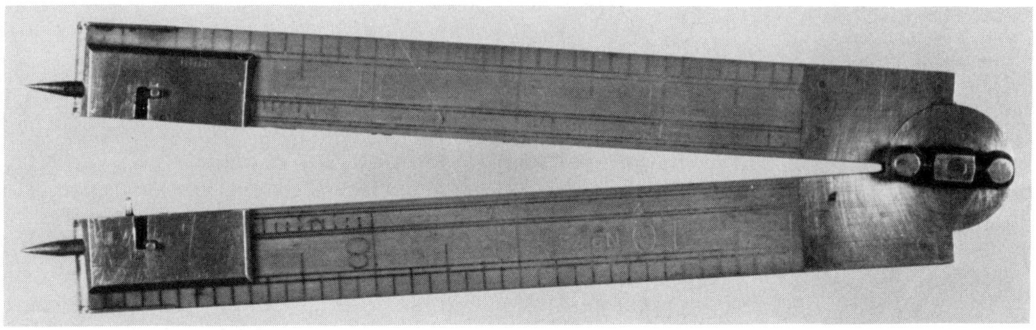

Figure 41: Nicol's Patent Rule Dividers (Rule Is a Stanley No. 72) *(Roger K. Smith Collection)*

Marsh's Dry Measure Gauge

This is an example of a Stanley rule regraduated by a second party. A Stanley No. 62½ has had elaborate gauging scales stamped into the brass binding on the rule edge (see Figures 42a and 42b), and then been remarketed under the graduator's name.

This modified rule was sold by T.E. Marsh, of Ware, Massachusetts, and dates from 1879–1921. It is not known if it was ever patented (the statement on the packing box notwithstanding).

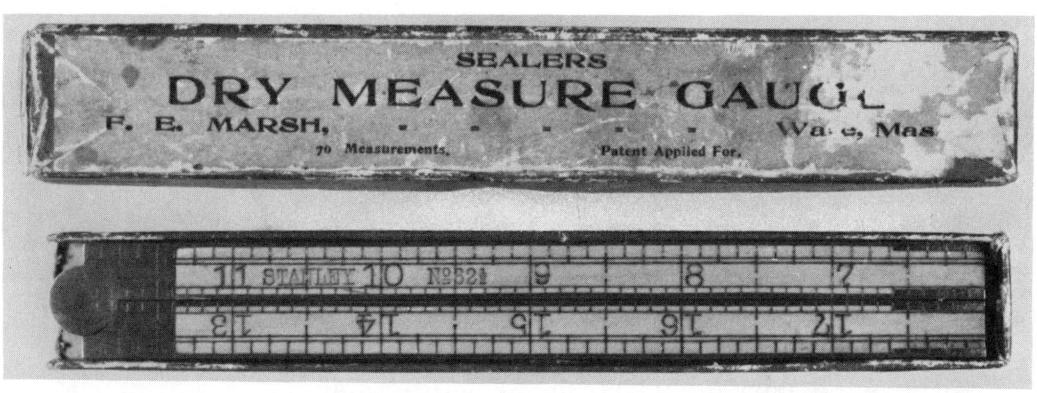

Figure 42a: Marsh's Dry Measure Gauge (Rule Is a Stanley No. 62½) *(Aime Blouin Collection)*

Figure 42b: Scales on Edge of Gauge *(Aime Blouin Collection)*

Rule Accessories

As a complement to their wide line of rules, Stanley also made and sold rule accessories for use with them, devices which could be attached to a rule to allow it to perform functions other than the nominal ones of marking straight lines and measuring. Three such accessories were made: The No. 1 Odd-Jobs Tool, the No. 2 3-Angle Tool, and the No. 99 Rule Trammel Points.

The No. 2 3-Angle Tool

The simplest of these was the No. 2 3-Angle Tool (see Figure 43). This device (which could also be used separately as a pocket level) attached to the rule in such a way that it (the rule) formed the blade of a square, mitre square, or marking gauge. Made of nickel-plated cast iron, it had a milled slot in one face into which the rule was clamped by a spring which pressed against one edge. When so attached to the rule, different surfaces of the tool formed fences at 30, 45, and 90 degrees to the edge, and allowed it to function as a common square, and as a mitre square for both four and six sided figures. This tool also allowed the rule to be used as a marking gauge; by sliding it along the rule until the right angle fence was the desired distance from the end, that end could then guide the pencil when the fence was held against the edge of the work.

That same right-angle fence, which incorporated a spirit vial, also functioned as the bearing surface when the 3-Angle Tool was used as a level. When used in this way, of course, the rule was not required.

The No. 2 3-Angle Rule Tool was introduced in 1912 (Ref. 27), and remained in the product line for twenty years, being offered for the last time in 1932 (Ref. 43).

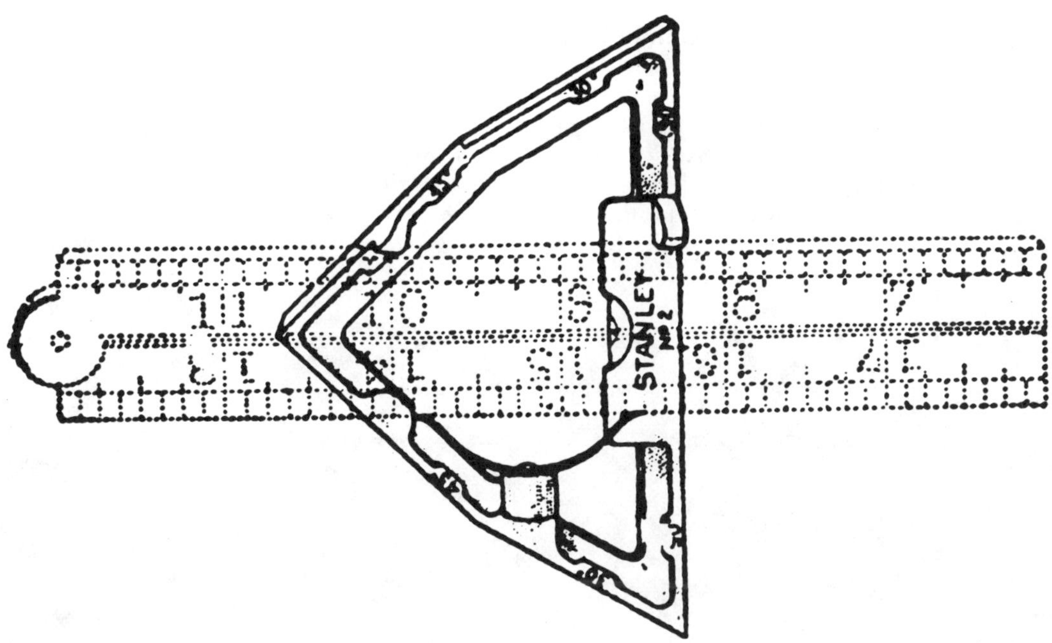

Figure 43: The Stanley No. 2 3-Angle Tool, 1914

(Ref. 29)

The No. 99 Rule Trammel Points

The second of these Stanley rule accessories were the No. 99 Rule Trammel Points (see Figure 44). These were attachments which used the rule as their stick or beam. They were not really much different from ordinary trammel points, such as the Stanley Nos. 1 through 6 (Ref. 25), the primary difference being that they were designed to fit a stick 3/16 inch thick (the usual thickness of a folding rule) and from 3/8 to 3/4 inch wide.

The trammel points consisted of a set of three brass heads, two with steel points, and a third with a pencil socket. These cylindrical heads were slotted to slip onto a rule edgewise, and threaded on the outside for thumb nuts which clamped them in place. Since they would fit rules of any width up to 3/4 inch, they could be used with all ordinary 2 fold and 4 fold rules. Only two of the heads would be used at a time. When employed as an ordinary trammel, the two with steel points were attached to the rule; when used as a beam compass, one was replaced with the third head containing the pencil socket.

Stanley offered these Rule Trammel Points for nearly forty-five years. They were listed for the first time in the 188 catalogue (Ref. 14), and remained in production until 1932 (Ref. 43), when, like so many other special purpose tools, they were dropped due to the exigencies of the Depression.

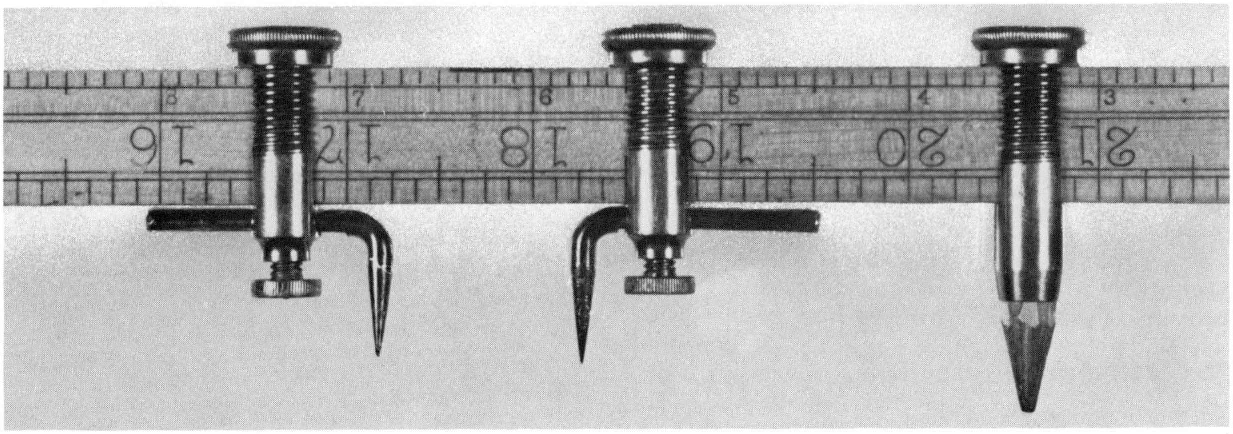

(*Charles and Walter Jacob Collection*)

Figure 44: Stanley No. 99 Rule Trammel Points

The No. 1 Odd-Jobs Tool

The third Stanley rule accessory was their *ne plus ultra,* the No. 1 Odd-Jobs Tool (see Figure 45), a masterpiece of multifunction design. Described in Stanley catalogues as "Ten Tools In One," the Odd-Jobs Tool on a 2 or 4 fold rule could be used as a marking or mortise gauge, a plumb or level, an inside or outside square, a mitre square, depth gauge, scratch awl, and beam compass.

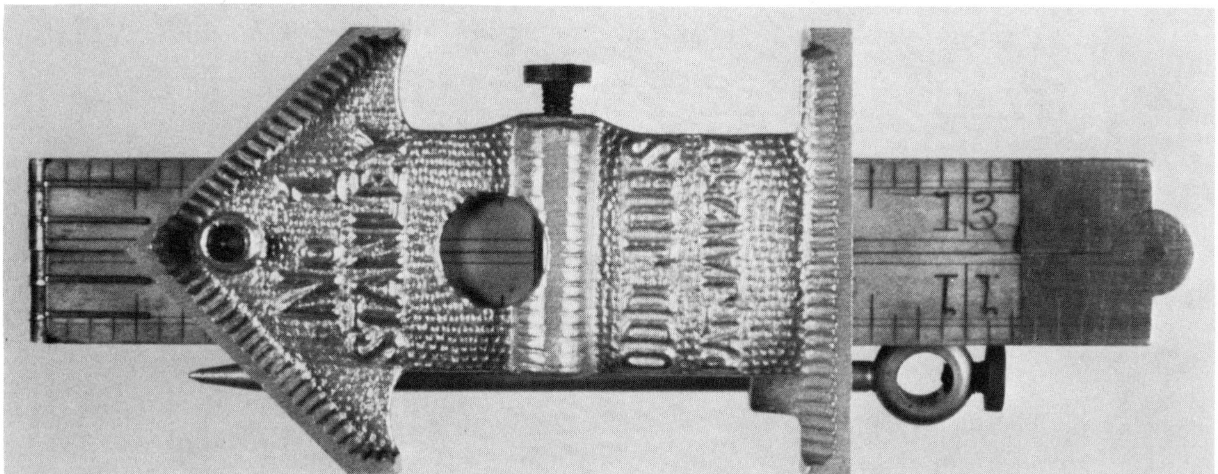

(*Eugene Frankio Collection*)

Figure 45: Stanley No. 1 Odd-Jobs Tool

Made of nickel-plated cast iron, the Odd-Jobs Tool was probably the most elaborate rule tool ever sold. Although in general form resembling the No. 2 3-Angle Tool and similar rule accessories from other manufacturers, it had a number of features which they did not.

The most notable was the ingenious scratch awl with which it was equipped. This awl, sharpened on one end, and provided with a pencil holder on the other, could be used independently, or clamped with a thumbscrew into the right angle face of the tool to act as the scribe of a marking or mortise gauge.

Equally important, the Odd-Jobs Tool was "stretched" to a length of nearly 4 inches, and the tips of the right angle and mitre fences milled parallel to the rule slot. This made it possible for the tool to also function as a plumb, using these milled tips as the bearing surface.

A third unique feature of this tool was the steel point located in one face near the junction of the two mitre fences. This point could be used as one end of a beam compass of which the other end was a pencil held by the scratch awl.

Stanley included elaborate diagrams and pictures (see Figure 46) in the catalogue to illustrate the many uses of this tool to the prospective purchaser. According to them (with only minimal exaggeration), this tool could enable "A Mechanic ... to do all ordinary jobs with only a Saw, a Hammer, and a Plane, in addition." (Ref. 18).

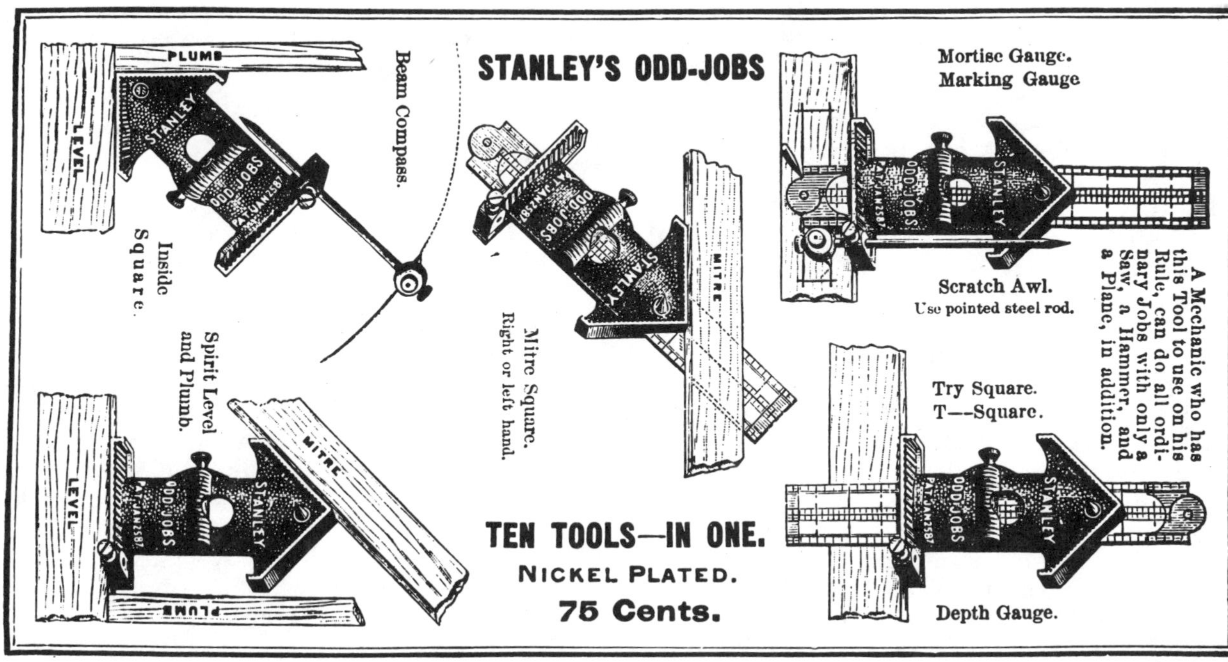

Figure 46: Applications of the Odd-Jobs Tool (Ref. 18)

The design of the Odd-Jobs Tool was based on a patent issued on Jan. 25, 1887, to George Hall, of Long Branch, New Jersey (U.S. Patent No. 356,533, entitled "T-Square and Gage Attachment for Rules"). Stanley must have purchased the patent from Hall very shortly thereafter, because the Odd-Jobs Tool first appeared in the 1888 catalogue (Ref. 14), only a year later.

Not all of the features in Hall's patent appeared in the final tool. The scribe, sliding in a hole in the right-angle fence, was retained, but Hall's complex scheme for clamping the rule at 45 degrees to the right angle fence for use as a mitre square was (probably wisely) abandoned and the Vee fence substituted.

Beginning in 1898, a rule specifically designed for use with the Odd-Jobs Tool was supplied with the tool. One foot long, and graduated in 16ths beginning ½ inch from the end, this rule was provided with a built-in trammel point located at the zero point of the scale. This rule/trammel point simplified the use of the Odd-Jobs as a marking gauge, and also extended its range when used as a trammel.

The Odd-Jobs Tool was produced for more than forty years, until, like the other two rule accessories, it was dropped from the catalogue in 1934 (Ref. 44).

Introduction

This chapter contains a listing of the wood, metal, and ivory folding rules manufactured by A. Stanley & Co., The Stanley Rule & Level Co., The Stanley Rule & Level Plant of the Stanley Works, and the Stanley Tools Division of the Stanley Works, from 1855 through 1975. Descriptive information is tabulated for each rule, and wherever possible one or more catalogue illustrations or photographs of an actual rule is included. Where appropriate, text has been included commenting on historical development, and describing model-specific features, peculiarities, and applications.

Rules are listed in numerical order, according to the rule number assigned to each in the Stanley catalogues, with the rules which were never assigned a number appearing first. Where a number was used for a rule, and then, after that rule was discontinued, subsequently reused for a different rule, the rules sharing this number are listed in chronological order.

Additional sections have also been included providing information on general types of rules (e.g., "BLINDMAN'S" RULES, BOARD MEASURE AND BOARD STICKS, etc.). These sections are located in the listing immediately prior to the description of the first rule of the type described.

Not all Stanley rules have been included. None of the Pull-Push rules are listed, nor are any of the Zig-Zag rules (unless one considers the No. 17 Blacksmiths' Rule as one of these). Tools which were not listed in the rule section of the catalogue, even if graduated, like many of the marking gauges, or provided with an accessory measuring stick, like the post-1894 Odd-Jobs Tool, have also been excluded. No rule is listed which has not appeared in at least one catalogue. These limitations have been imposed for two reasons: to maintain the focus of this study on those rules which are of primary historical interest, and to prevent its growth to the point where its publication would become prohibitively expensive.

The Stearns folding rules, which were produced by Stanley between 1863 and 1898, have also been excluded. These rules were quality products, and the early history of E. A. Stearns & Co. is undoubtedly worthy of research and description, but after the purchase of that company by Stanley in 1863, this product line ceased to grow and evolve. No new rules bearing the Stearns name were introduced, and one by one during this 35 year period the existing rules were canceled. It was felt that the history of these rules could best be left for a separate study, at some future date, which would cover the entire history of the line from 1833 on.

If the reader feels that these decisions, as to which rules to include and which to omit, are wrong, the author can only reply that the choice seems right to him, but that he would be happy to discuss the subject at any future time.

Long's Gear Calculating Rule, offered by Stanley in the 1870 catalogue (Ref. 8), was intended for use by designers and machinists to simplify the problems of designing and machining gears.

Invented by Charles B. Long, of Worcester, Massachusetts, and patented by him on April 25, 1865 (U.S. Patent No. 47,436, entitled "Gear Cutting Rule"), it was intended to provide most of the data needed by designers, patternmakers, and machinists in laying out the teeth on gears prior to making a pattern, or cutting the teeth on a gear casting.

Long's rule was a 2 foot, 2 fold rule, constructed similarly to a patternmakers' folding rule (that is, with a knuckle joint instead of the usual rule joint), and was brass bound. Its surfaces were marked with elaborate scales, inside and out, with graduations of 10ths, 12ths, 14ths, and 16ths of inches along the four edges, and tables for eight different tooth spacings down the surface in between. These tables had been calculated to include allowances for factors such as pitch line, etc., and thus could be used by a designer or machinist without many of the elaborate calculations usually accompanying gear design.

CHAPTER 5 The Rules and Their Features

(No #) Gear Calculating Rule, 2 Foot, 2 Fold

Offered:	1870
Construction:	Knuckle Joint, Full Bound
Material:	Boxwood, Brass Hinge & Trim
Width:	1³/₈ Inch
Graduations:	10ths, 12ths, 14ths, and 16ths of Inches
Other Markings:	Gear and Cog Wheel Tables

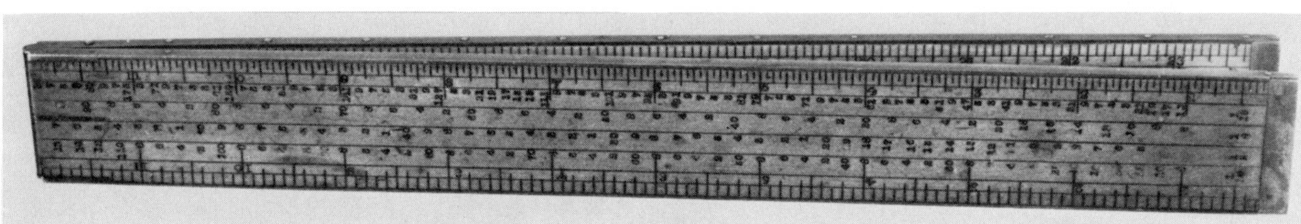

Figure 47a: Gear Calculating Rule, 1870 *(Paul Kebabian Collection)*

The description of the Gear Calculating rule in the catalogue was unusually fulsome, and was accompanied by testimonials from various eminent machinists and machine shops, but in spite of this it was apparently not successful. By 1872 (Ref. 9) it had been removed from the catalogue, and was never offered again.

Figure 47b: Tables on the Gear Calculating Rule *(Ref. 8)*

(No #) Carpenters' ("Four-Square") Rule, 2 Foot, 4 Fold
Offered: 1926 thru 1930
Construction: Round Joint, Middle Plates
Material: Boxwood, Brass Hinges & Tips
Width: 1 Inch
Graduations: 8ths and 16ths of Inches

Stanley's so-called "Four-Square" rule was one of a whole line of inexpensive tools which was offered from 1926 to 1934 specifically aimed at the homeowner or occasional woodworker. In actuality, this rule was the Stanley bottom of the line No. 68A 4 fold rule, with the Four-Square trademark substituted for the rule number (the Four-Square rule had no number), and packaged in a printed cardboard sleeve.

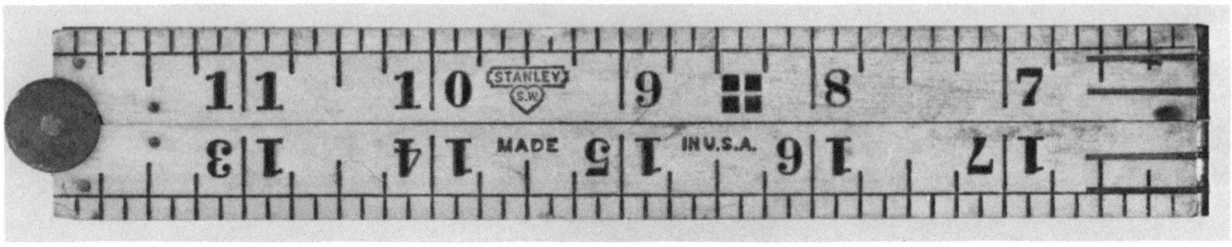

Figure 48a: Four-Square Rule, 1926—1930 *(Charles and Walter Jacob Collection)*

Figure 48b: Four-Square Rule Advertising Display *(Charles and Walter Jacob Collection)*

The Four-Square rule was "introduced" in 1926 (Ref. 39), at the same time as the rest of this line of tools, and like all of them was promoted with elaborate advertising in the press and at the point of sale (see Fig. 48b). It was discontinued in 1932 (Ref. 43); apparently it just was not worth while marketing the No. 68A under two different identities, particularly in the face of the worsening economic depression of 1929–1939.

036 Combination Rule, 1 Foot, 2 Fold

Offered:	1929 thru 1941	
Construction:	Square Joint, Full Bound, 4½ Inch Blade In one Leg, Spirit Level In Other Leg	
Material:	Boxwood, Brass Hinge & Trim, Steel Blade	
Width:	1⅜ Inch	
Thickness:	⅜ Inch	
Graduations:	Body:	10ths, 12ths, 16ths, and 24ths of Inches
	Blade:	24ths of Inches
Scales:	Body:	Drafting (¼"=1' & ½"=1')
		Degrees (20 to 75) (on Inside Edge)
	Blade:	Degrees (0 to 45)
		Pitch to the ½ Foot

The No. 036 combination rule is the most universal measuring tool ever sold by Stanley. It could be used as a rule, protractor, plumb, level, square, bevel gauge, or inclinometer, and was rivaled in its versatility only by the Stanley No. 1 odd-jobs tool

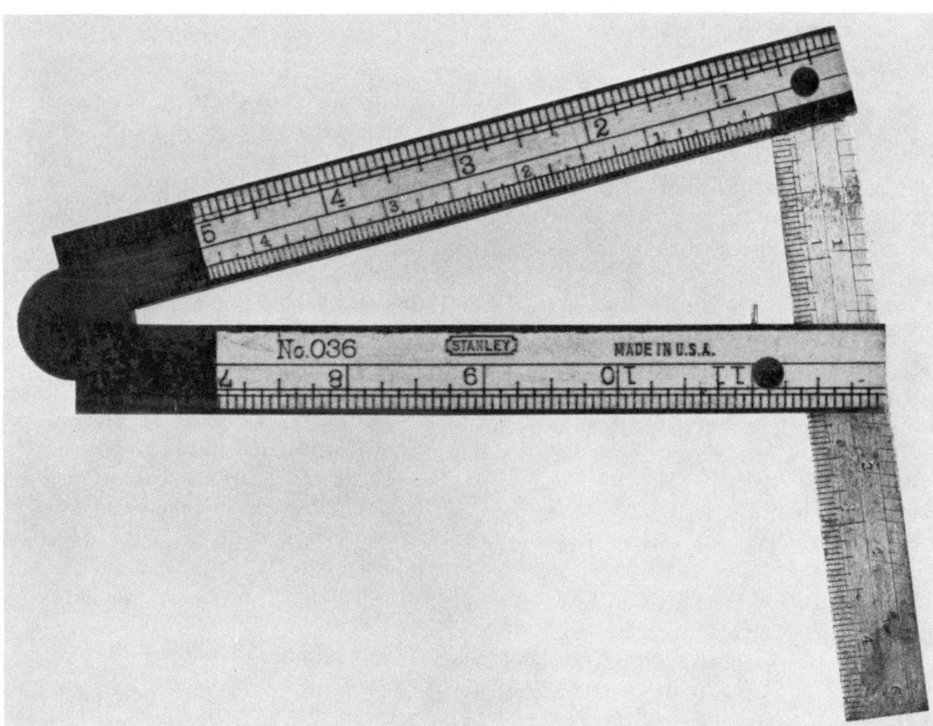

Figure 49a: No. 036 Combination Rule, 1934–1941 *(Ken Roberts Collection)*

In form, the combination rule was a simple device: a 1 foot, 2 fold rule with a 4½ inch blade on the inside of one leg, and a spirit level on the outside of the other. A small slot and clamping screw in the tip of this second leg allowed it to be clamped to the blade at any point, thus making it possible to lock the rule partially open at any angle between 0 and 45 degrees.

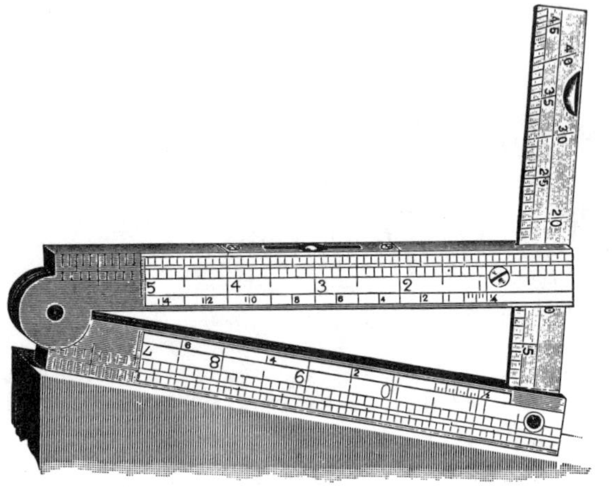

(Ref. 30)
Figure 49b: No. 036 As An Inclonometer

It was the graduations on the legs and blade, however, which rendered the combination rule so versatile and useful. The blade was calibrated in both degrees and pitch to the ½ foot, to facilitate use as an inclinometer, and the body, in addition to the usual scales of 8ths, 10ths, 12ths, and 16ths of inches, was also graduated in 24ths of an inch, and marked with Drafting Scales and another degree scale.

By proper use of the various scales and features of the rule, it was possible to use the rule in any of the following ways:

- As a 1 foot, 2 fold rule
- As a plumb or level
- As a try square or bevel gauge
- To measure slope in degrees or pitch to the foot
- To scale braces (as in framing buildings)
- As a drafting rule (on scale drawings)
- To lay out angles
- To measure the height of inaccessible objects

The No. 036 combination rule was not originally a Stanley product, but was the continuation, under the Stanley name, of a rule which had been manufactured prior to 1929 by the Chapin-Stephens Co., manufacturers of planes, rules, gauges, and other woodworking tools.

The combination rule was originally invented by Lorenzo Stephens, founder of the rule-making firm of L.C. Stephens & Co., and was patented by him on Jan. 12, 1858. Manufactured in both boxwood and brass (Stephens' No. 36) and ivory and German silver (Stephens' No. 38), this rule was for many years the leader of the extensive Stephens line, and was sold at a premium price.

In 1901, that company, by then known as D.H. Stephens & Co., was merged with the competing rule- and plane-making firm of H. Chapin's Son & Co. to form the Chapin-Stephens Co. All of the D.H. Stephens Co. line of rules, including the No. 36 (renumbered as the No. 036), were continued in production by this successor firm.

In 1927, the Chapin-Stephens Co., by then severely pressed financially by the shrinking market for wood planes and folding rules, was reorganized, and two years later, in 1929, finally ceased production. At that time the Stanley Rule & Level Division of the Stanley Works acquired the line of rules from the defunct company. Of all the rules so purchased, only the No. 036 was continued in production under the Stanley name as a unique, identifiable product, listed in the catalogue. Presumably, all the other rules were, if unfinished, completed and marked as Stanley rules, or, if already stamped with the Chapin-Stephens trademark, sold under that name until the stocks were exhausted.

Stanley produced the No. 036 only in boxwood and brass, having discontinued all production of ivory rules prior to about 1920. In the large dealers' catalogues (Refs. 44, 46, and 47 it was described as a "Miners' Combination Rule"; in the smaller No. 34 catalogues (Refs. 42, 43, 45, and 48 through 51) simply as a "Combination Rule." The rule continued to be part of the product line until 1941, when, as in the case of many other rules, it was eliminated in order to make room for higher priority war production.

1 Carpenters' Rule, 2 Foot, 2 Fold
Offered: 1855 thru 1917
Construction: Arch Joint
Material: Boxwood, Brass Hinge & Tips
Width: 1½ Inch
Graduations: 8ths and 16ths of Inches
Scales: Octagonal

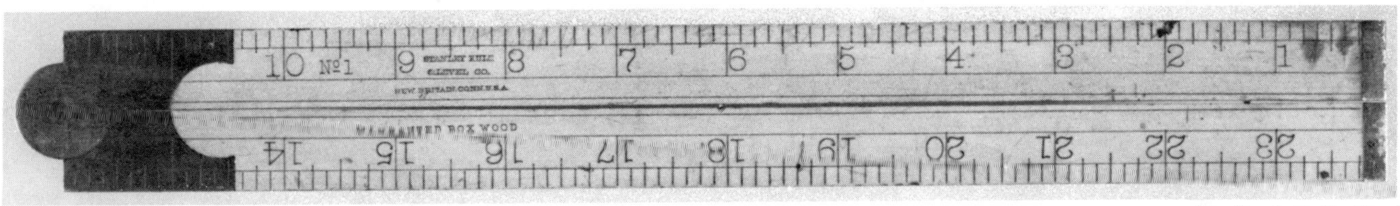

Figure 50: No. 1, 1858–1917 *(Author's Collection)*

2 Carpenters' Rule, 2 Foot, 2 Fold
Offered: 1855 thru 1905
Construction: Arch Joint, Bitted
Material: Boxwood, Brass Hinge & Trim
Width: 1½ Inch
Graduations: 8ths, 10ths, and 12ths of Inches (1862 and before)
8ths, 10ths, and 16ths of Inches (1865 and after)
Scales: Octagonal (1860 and after)

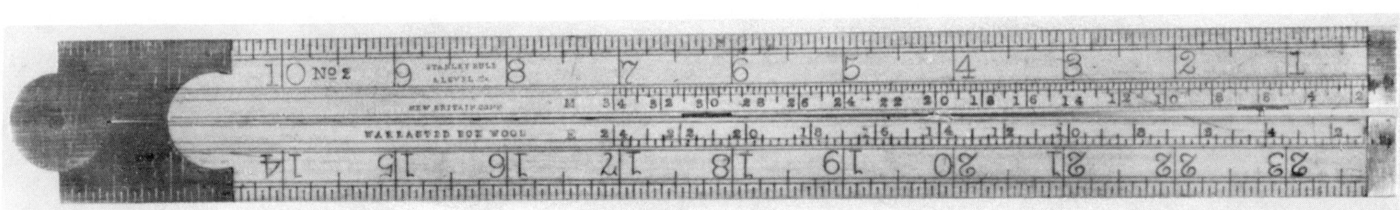

Figure 51: No. 2, 1858–1905 *(Charles and Walter Jacob Collection)*

3 Carpenters' Rule, 2 Foot, 2 Fold
Offered: 1855 thru 1859
Construction: Arch Joint, Half Bound
Material: Boxwood, Brass Hinge & Trim
Width: Not Known, but probably 1½ Inch
Graduations: Not Known

Half bound 2 foot, 2 fold rules were only part of the Stanley product line for a very short time. Only two were made, the No. 3 (without slide) and the No. 14 (with slide), both offered from 1855 (Ref. 1) through 1859 (Ref. 2).

It is not known why Stanley discontinued these two rules after only a short period. Other half bound rules, of the 2 foot, 4 fold type, were apparently well received, and remained in the product line for many years (more than 70 years in the case of the No. 84).

It is possible that these two rules were discontinued due to problems with warping. Binding only one edge of a piece of boxwood with brass could cause unbalanced stress as the wood expands and contracts with varying humidity, and the stresses would be more severe on a 1 foot piece than on one only half as long.

3 Carpenters' Caliper Rule, 1 Foot, 4 Fold

Offered:	1900 thru 1917	
Construction:	Square Joint, Full Bound, Left Hand Caliper	
Material:	Boxwood, Brass Hinges, Caliper & Trim	
Width:	5/8 Inch	
Graduations:	Body:	8ths and 16ths of Inches
	Caliper:	16ths and 32nds of Inches (1900) 16ths of Inches (1902 and after)

Figure 52: No. 3, 1900–1917 *(Roger K. Smith Collection)*

4 Carpenters' Rule, 2 Foot, 2 Fold

Offered:	1859 thru 1917
Construction:	Arch Joint (plates on outside of wood), Bitted
Material:	Boxwood, Brass Hinge & Trim
Width:	1½ Inch
Thickness:	"Extra Thin" (1/10 Inch)
Graduations:	8ths and 16ths of Inches (1902 and before) 8ths, 10ths, and 16ths of Inches (1909 and after)
Scales:	Drafting Octagonal

The No. 4 is unique among Stanley rules, in that the plates for the rule joint are mounted on, instead of being set into, the surface of the wood as on all other hinged rules. This was necessary due to the extra thinness of the wood; if the plates had been let into the wood in the ordinary way, hardly any wood would have remained between them, leaving the joint dangerously weakened.

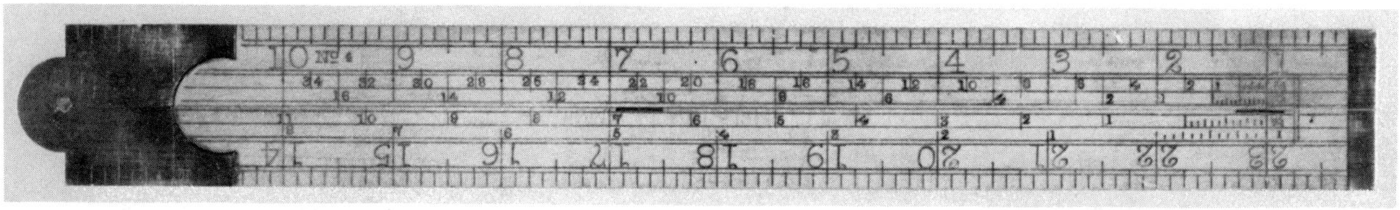

Figure 53a: No. 4, 1858–1917 *(Charles and Walter Jacob Collection)*

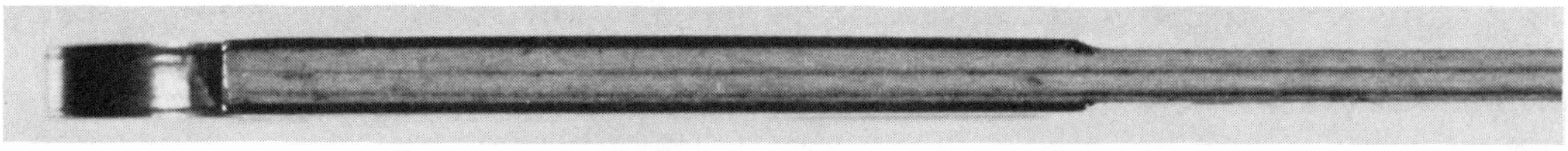

Figure 53b: Edge View of the No. 4 Joint *(Charles and Walter Jacob Collection)*

In the case of the No. 4, the bitting was not an extra feature, but a necessity; the wood was so thin that the alignment pins would have broken out very quickly.

Although the fitting of the joint during manufacture must have been easier for the No. 4, due to the lack of any need to rabbet the stock for the plates, the cost was higher than a comparable rule of normal thickness (such as the No. 2). In 1898, for example, the No. 2 was priced at $8.00 per dozen, while the No. 4 was priced at $10.00 per dozen (Ref. 17). The difference was probably due to the extra care required in preparing the thin wood stock, and in graduating the rule with the plates protruding above the surface.

5 Carpenters' Rule, 2 Foot, 2 Fold
Offered: 1855 thru 1932
Construction: Arch Joint, Full Bound
Material: Boxwood, Brass Hinge & Trim
Width: 1½ Inch
Graduations: 8ths, 10ths, and 16ths of Inches
Scales: Drafting
Octagonal

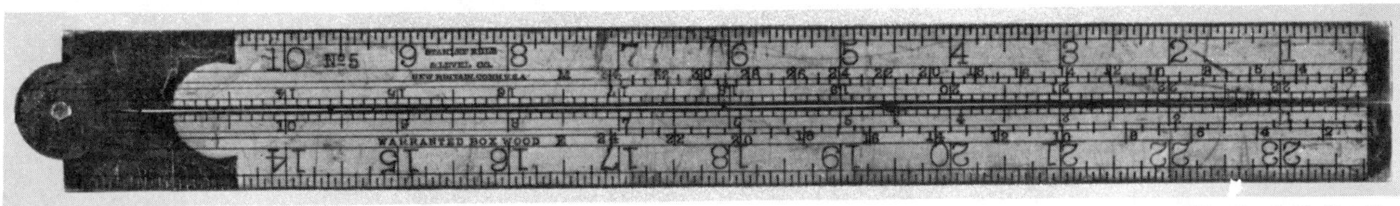

(Author's Collection)

Figure 54: No. 5, 1858–1921

The 1932 catalogue (Ref. 43) was not the last listing of the No. 5. This rule was included in the December, 1942, and November, 1943, corrections to the July, 1941, dealers' price list (Refs. 52 and 53) (and thus, by implication, in that 1941 list as well). This was only for the purpose of disposing of existing stock, however, and did not represent a resumption of production. It is not known whether these remaining No. 5's were marked with the "Sweetheart" trademark characteristic of the early 1930's, or with the "Made IN USA" trademark of the post-1932 period, as was the case with other rules in the same situation.

The Engineers' (Sliding) Rule

Unquestionably the most elaborately marked and graduated Stanley rules were the No. 6 and No. 16 engineers' rules manufactured between 1855 and 1902–1905.

Invented in the early nineteenth century by Josiah Routledge of Boulton, England, these rules were an attempt to provide the user with not only the means for performing rapid calculations (the Gunter's slide), but also much of the physical data required to work out common problems. With an engineers' rule, a user could perform volume conversions, weight computations, geometric analysis, and steam engine/pump computations (this last was a significant class of problems 150 years ago; from its inception, the steam engine had been used to pump water from mines, etc., and this was still one of its most common uses).

Physically, the engineers' rule was a 2 foot, 2 fold rule with slide, the No. 6 having the same construction as the No. 12, and the full bound No. 16 being similar to the No. 15. The rules differed from their non-engineering counterparts in two respects: the drafting scales were omitted, to be replaced by tables of physical and geometrical data, and "Gauge Points," and the D scale of the Gunter's slide was marked differently from that on the carpenters' (sliding) rule.

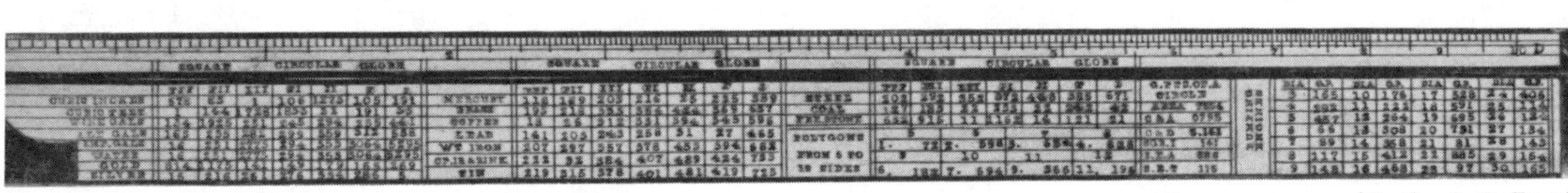

Figure 55: Tables on the Engineers' Rule

(Author's Collection)

Using the data on the rule, and the Gunter's slide, the skilled user of the engineers' rule could solve whole classes of problems in mensuration, engineering, and maining without recourse to any other aids or tables.

a) The Tables:

The tables on the engineers' rule were five in number, and contained reference data organized as follows:

TABLE 1: Conversion factors for relating the volumes of various geometric solids (square [rectangular parallelopipeds], circular [cylinders], and globe [spheres]), to the various units of volume (square inches, wine gallons, etc.). The relationships are expressed in reciprocal form (e.g., a one foot cube (FFF) contains 1728 cubic inches; 1/1728 = .0005787 ...; hence 578).

TABLE 2: Conversion factors for relating the volumes of the same geometric solids as in Table 1 to their weight in pounds for various materials. Again, the relationship is expressed in reciprocal form (e.g., a one foot long, one inch diameter cylinder (FI) of water weighs 0.3403 ... pounds; 1/0.3403 ... = 2.938 ...; hence 294).

TABLE 3: A table of the areas (in square units) of regular polygons (pentagon, hexagon, etc.) of unit side, for polygons from five to twelve sides. (e.g., an octagon with 1 foot sides has an area of 4.828 ... square feet)

Tables 1, 2, & 3 occupy the three leftmost areas of the leg, but are run together (as is shown in the figure above) and then arbitrarily separated into sections with equal numbers of rows (at the points indicated by the asterisk [*]). Thus the first area contains Table 1 and part of Table 2, the second, more of Table 2, and the third, the rest of Table 2 and Table 3.

The values in Tables 1 & 2 are expressed in reciprocal form in order to permit the simplest possible operations on the Gunter's slide. This has the double advantage of reducing the cumulative errors which accrue as multiple-step operations are performed on the slide, and of making the sequence of steps required to be learned by the user as simple as possible.

TABLE 4: Gauge Points of a Circle. A table of numbers relating the diameter, area, an and circumference of a circle, and further relating them to the dimensions of the square and triangle which can be inscribed therein.

The significance of the various terms is as follows:

AREA: The area of a circle of unit diameter (Actual value: 0.78539 ...)

C & A: The area of a circle of unit circumference (Actual value: 0.07958 ...)

C & D: The circumference of a circle of unit diameter (This is Pi; actual value: 3.1416 ...)

SQR. I: The diameter of a circle within which is inscribed a square of unit side (Actual value: 1.414 ...)

S.E.A.: The side of a square equal in area to a circle of unit diameter (Actual value: 0.8862 ...)

S.E.T.: The diameter of a circle within which is inscribed an equilateral triangle of unit side (Actual value: 1.155 ...)

TABLE 5: Gauge Points (G.P.) for Pumping Engines. A table of values for computing the required cylinder diameter for a steam engine driving a pump of known cylinder diameter which is raising water a known height. This table presupposes steam pressure to operate the engine of 7 pounds per square inch, and that the cranks on the engine and the pump have the same swing.

The gauge points are calculated for a series of pump diameters (DIA), such that the square root of the product of the gauge point and the height is the diameter required for the pumping engine cylinder (e.g., in order to drive a 6 inch pump which raises water 4½ yards, a pumping engine run by 7 psi steam would be required to have a cylinder diameter of $D = \sqrt{6.60 \times 4.5} = \sqrt{29.7} = 5.45$ inches).

In Tables 1, 2, 4, & 5, the various values are expressed without showing the location of the decimal point. The slide rule does not (and cannot easily) keep track of decimal points during manipulation; ordinarily it is left for the user to do this separately, or to scale the result based on a priori knowledge. Since including the decimal point thus contributed nothing to reaching the final solution, and only made the stamped values occupy more space, it was omitted in the interest of compactness.

		SQUARE			CIRCULAR		GLOBE	
		FFF	FII	III	FI	II	F	I
TABLE (1)	CUBIC INCHES	578	83	1	106	1273	1105	191
	CUBIC FEET	1	144	1728	1833	22	191	33
	WINE GALS	134	1925	231	245	294	255	441
	ALE GALS	163	235	282	299	359	312	538
	IMP. GALS	16	231	2773	294	353	3064	5295
TABLE (2)	WATER	16	231	2773	294	353	3064	5295
	GOLD	814	1175	141	149	179	155	269
	SILVER	15	216	261	276	334	286	5
	* MERCURY	118	169	203	216	258	225	389
	BRASS	193	278	333	354	424	369	637
	COPPER	18	26	312	331	394	344	596
	LEAD	141	203	243	258	31	27	465
	WT IRON	207	297	357	378	453	394	682
	CP. IR & ZINK	222	32	384	407	489	424	733
	TIN	219	315	378	401	481	419	723
	* STEEL	202	292	352	372	448	385	671
	COAL	127	183	22	233	280	242	42
	FRE STONE	632	915	11	1162	14	121	21

TABLE (3)					
	POLYGONS FROM 5 TO 12 SIDES	5	6	7	8
		1.72	2.598	3.634	4.828
		9	10	11	12
		6.182	7.694	9.366	11.196

Figure 56: Tables 1 (Volume Conversion Factors), 2 (Weight Calculation Factors), and 3 (Polygon Areas)

G. PTS. OF A		DIA	G.P	DIA	G.P.	DIA	G.P.	DIA	G.P.
CIRCLE	P E	3	165	10	183	17	528	24	106
AREA 7854	U N	4	292	11	222	18	591	25	114
C & A 0795	M G	5	457	12	264	19	661	26	124
C & D 3.141	P I	6	66	13	308	20	731	27	134
SQR.I 141	I N	7	89	14	358	21	81	28	143
S.E.A. 886	N E	8	117	15	412	22	885	29	154
S.E.T. 115	G S	9	148	16	468	23	97	30	165

|← TABLE (4) →|← TABLE (5) →|

Figure 57: Tables 4 (Gauge Points of a Circle), and 5 (Pumping Engine Gauge Points)

b) The Gunter's Slide

The Gunter's slide on the engineers' sliding rule was identical with that on the carpenters' sliding rule (Coggeshall's pattern) except for one scale, the D scale.

On the carpenters' rule this scale, also labeled the GIRT LINE, was "folded" at 4; that is, it began at the value 4 at the left end, and progressed through 5, 6, etc., to 1 near the middle, and thence through 2, 3, etc., to 4 again at the right.

On the engineers' rule the D scale was not folded; that is, it began at the value 1 at the left end, and progressed through 2, 3, etc., to 1 again on the right.

The purpose of folding a scale on a slide rule is to reduce the number of manipulations required in common operations. Folding accomplishes this in two ways. First, a folded scale introduces the folding point as a multiplying or dividing factor when transferring points between the folded scale and other scales during computations; a scale with the proper folding point can thus in many cases reduce the number of slide settings by one or more. Second, folding a scale has the effect of moving numbers near $^1/_{10}$ toward the physical center of the scale. The physical location on the scales of the numbers involved in a computation will often influence the number of manipulations necessary to perform it; by moving a particular number near the center it is often possible to reduce the number of steps for a particular type of problem by one.

Coggeshall had recognized this when he developed his version of the slide rule. Since he was particularly interested in timber measurement, and one of the constants in his log rule was 12 (inches in a foot), he folded the D scale on his rule at 4, to move the value 12 almost exactly to the center.

Routledge could not use a folded D scale on the Gunter's slide of his rule. In computing the values in his various tables all constant factors (such as 4 or Pi) other than the geometric dimensions specific to the particular problem had already been included; a folded scale would have introduced an unwanted factor and rendered the answer useless. He could have compensated for this by computing the table values to include this folding point factor, but this would have effectively obscured their physical significance.

6 Engineers' Sliding Rule, 2 Foot, 2 Fold

Offered:	1859 thru 1905	
Construction:	Arch Joint, Slide in One Leg Bitted (1860 and after)	
Material:	Boxwood, Brass Hinge, Slide, & Trim	
Width:	1½ Inch	
Graduations:	Body:	8ths, 10ths, and 16ths of Inches 100ths of a Foot (On Outside Edge)
	Slide:	8ths of Inches (On Back)
Scales:	Body:	Gunter's, 2 Cycles Gunter's, 1 Cycle Octagonal
	Slide:	Gunter's, 2 Cycles
Other Markings:	Volume, Weight, Polygon, Circle, & Pumping Engine Tables	

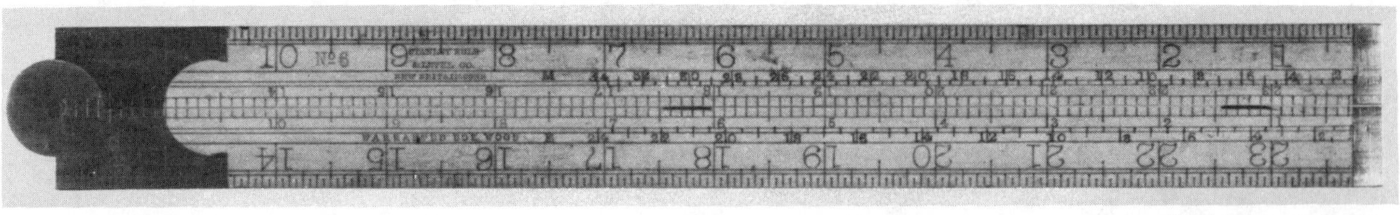

(Bud Steere Collection)

Figure 58a: No. 6, 1859–1905 (Front)

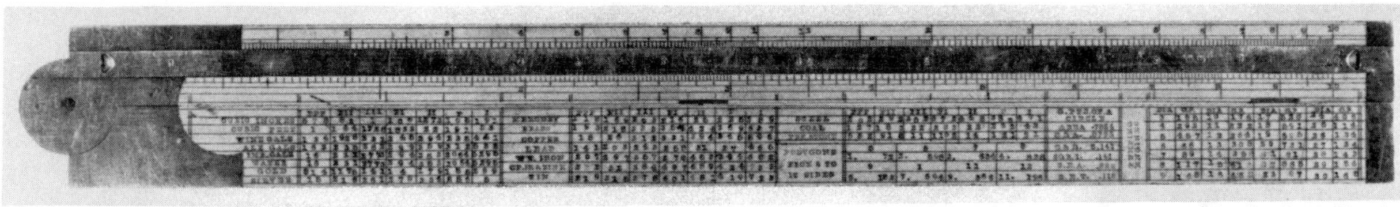

Figure 58b: No. 6, 1859–1905 (Back)

(Author's Collection)

7 Carpenters' Rule, 2 Foot, 2 Fold

Offered: 1855 thru 1859
Construction: Arch Joint
Material: Boxwood, Brass Hinge & Tips
Width: Not Known, but probably 1½ Inch
Graduations: Not Known
Other Markings: Board Measure Tables

"Blindman's" Rules

Prior to 1900, the graduations and figures on Stanley rules were always put on by the same method; by scribing or stamping them into the wood, and then filling the resulting incisions with black paint or wax. The resulting markings were thin, clear, and as durable as the surface of the wood itself. The only disadvantage of this method was that the markings were not as dark and heavy as some would have; they were hard to read if the light was dim, or if the user had impaired vision.

In 1900, partly as a response to this problem, and also partly as a result of improvements in ink/printing technology, Stanley began to manufacture and sell their so-called "Blindman's" rules. The markings on these rules were printed on, instead of being incised into, the surface of the wood, and thus could be made extra wide and extra dark. Additionally, the figures were made almost twice as large as on ordinary rules, improving their readability even further.

Only one Blindman's rule, the 2 foot, 4 fold No. 7 (Ref. 18), was initially offered. This rule was immediately popular, and was to remain in the product line for many years, being only discontinued between 1950 and 1955 as one of the last remaining 4 fold rules.

A second Blindman's rule, the 3 foot, 4 fold No. 8 was added to the line in 1909 (Ref. 25), but it was apparently not as successful

Figure 59: Regular (incised) and Blindman's (printed) Figures

(Author's Collection)

as the No. 7, and was discontinued after only 8 years, in 1917 (Ref. 32).

In 1929 (Ref. 41), another 3 foot, 4 fold rule with Blindman's markings was offered. This new rule, the No. 170BE, more nearly resembled the No. 7 than the No. 8, being 1-3/8 inch wide, and having middle plates. As was the case with the No. 8 in 1909–1917, the No. 170BE was only moderately well received, and remained in the catalogue for only a few years, until 1938.

All of these rules, the Nos. 7, 8, and 170BE, were graduated left to right, instead of right to left as was the norm with American makers. The Nos. 7 and 8 were the first Stanley 4 fold rules to be offered this way. The reason for this departure from the standard scheme is not known; no explanation is given in the catalogue, nor in any of the literature examined by the author. By the time the No. 170BE was introduced in 1929, rules with left to right ("English") graduations had become more common, and it is less surprising that this rule should have been so marked.

7 Carpenters' ("Blindman's") Rule, 2 Foot, 4 Fold

Offered:	1900 thru 1950
Construction:	Square Joint, Edge Plates (1932 and before) Square Joint, Middle Plates (1934 and after) Extra Large & Heavy Figures, and Extra Heavy Graduations Left-to-Right ("English") Markings
Material:	Boxwood, Brass Hinges & Tips
Width:	1³/₈ Inch
Graduations:	8ths and 16ths of Inches
Scales:	Drafting (1941 and after)

The No. 7, introduced in 1900 (Ref. 18), was the first of the so-called "Blindman's" rules, with extra large and heavy figures and extra heavy graduations. Like the other rules of this type, the Nos. 8, and 170BE, it was graduated from left to right instead of from right to left as were most other 4 fold rules.

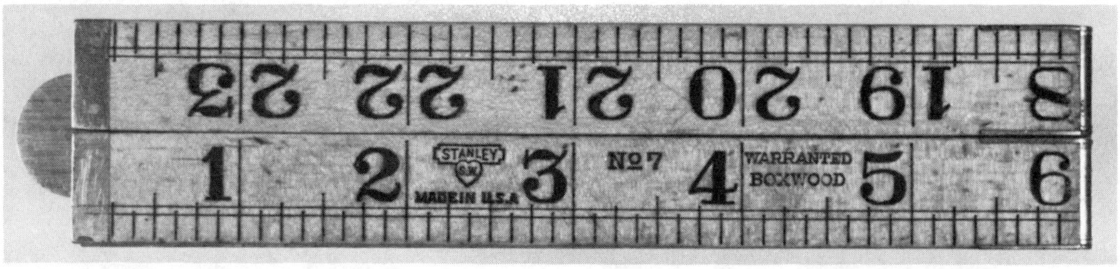

Figure 60a: No. 7, 1922–1932

(Author's Collection)

In 1934 the No. 7 was changed from having edge plates (see Figure 60a) to having middle plates (see Figure 60b). This was probably a cost reduction measure. By the early 1930's Stanley was only producing three 4 fold rules 1³/₈ inch wide. One was the No. 7; the other two, the Nos. 70 and 170BE, both had middle plates; by changing the No. 7 to conform, the need to manufacture and fit edge plates to 1³/₈ inch rules was eliminated.

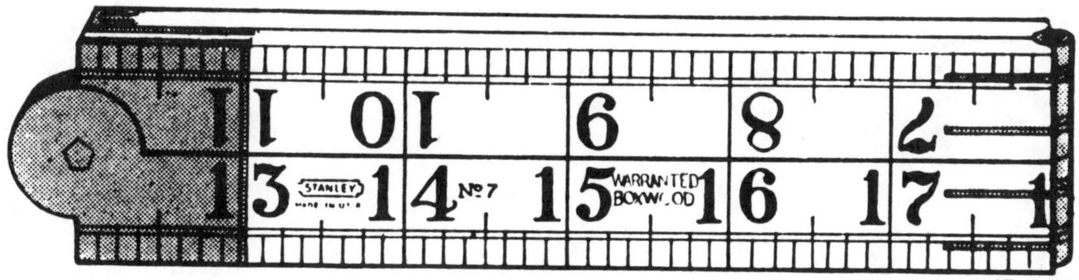

Figure 60b: No. 7, 1934 (Ref. 44)

A number of these rules have been observed which are marked both No. 7 and No. 70E. These must date from the period 1941–1943, after the No. 7 had drafting scales added to its inside surface. With this change, the only remaining difference between the No. 7 and the No. 70E, introduced in 1937, was their style of marking (the No. 70E had heavy, instead of blindman's figures). Apparently when this change was made in the No. 7 it was decided to merge the two rules, selling the No. 7 as a No. 70E as well. Two years later, when the No. 70E was discontinued, this dual marking scheme was dropped, and the No. 7 resumed its previous single identity.

The No. 7 was one of the most widely accepted of the Stanley 4 fold rules, and remained in the product line for more than 50 years. It was one of the 17 rules to remain in production after World War II, and was only discontinued in 1955 as these few rules were phased out one by one.

8 Carpenters' Rule, 2 Foot, 4 Fold

Offered:	1859 thru 1888
Construction:	Round Joint, Middle Plates
Material:	Boxwood, Brass Hinges & Tips
Width:	1 Inch
Thickness:	"Extra Thick"
Graduations:	8ths and 16ths of Inches

8 Carpenters' ("Blindman's") Rule, 3 Foot, 4 Fold

Offered:	1909 thru 1917
Construction:	Square Joint, Edge Plates Extra Large & Heavy Figures and Extra Heavy Graduations Left-to-Right ("English") Markings
Material:	Boxwood, Brass Hinges & Tips
Width:	1 Inch
Graduations:	8ths and 16ths of Inches

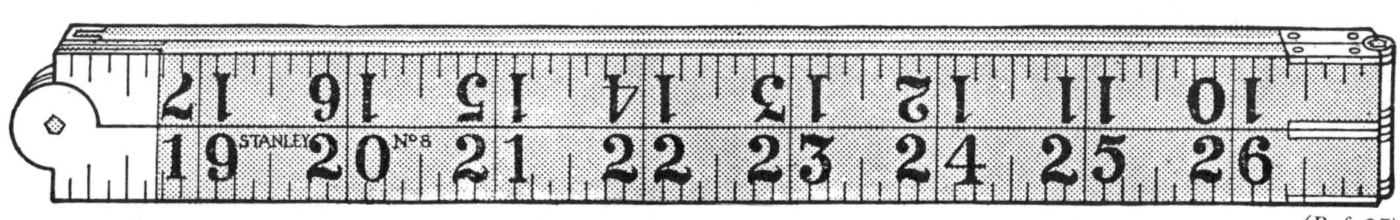

Figure 61: No. 8, 1912 (Ref. 27)

The No. 8, introduced in 1909 (Ref. 25), was the second of the so-called "Blindman's" rules, with extra large and heavy figures and extra heavy graduations. Like the other rules of this type, the Nos. 7 and 170BE, it was graduated from left to right instead of from right to left as were most other 4 fold rules.

Apparently the No. 8 was not extremely popular, or perhaps the period just prior to the First World War, with competition from

Zig-Zag rules increasing, was a bad time to be introducing new rules. At any rate, this rule was discontinued after 1917, after being in production for only eight years.

Metric Rules

The Stanley Company has manufactured and marketed rules with metric graduations since at least before 1867, and as late as 1943. During two periods, 1877 thru 1879, and 1919 thru 1943, such rules were standard products, listed by number in the catalogue. However, metric rules had been offered as special items, to order, as early as 1867, and continued to be offered as such, in parallel with the standard metric rules, as late as 1929. Prior to 1909, it is not indicated in the catalogue which models of rules could be supplied with metric graduations, just that "With recently constructed machinery, we can furnish, to order, Rules marked with Spanish graduations, or with Metric graduations" (Ref. 13). In 1909, the availability of special order metric rules was expanded to include any standard rule, with the exception of the shrinkage and extension rules, and in 1912 further still to include full metric (both sides) or metric/inch (one side only).

The first marketing of metric rules as standard products was in 1877 (Ref. 11). Two types of metric rules were offered: rules in metric lengths, and rules in 1 and 2 foot lengths. Both types were marked with both inch and metric graduations, from left to right. The first group consisted of the Nos. 10 (50 cm., 4 fold), 20 (1 meter, 4 fold), and 30 (1 meter, 4 fold). The second group consisted of the Nos. 101, 151, 161, 165, and 173, where each was the metric/inch equivalent of the corresponding standard rule numbered 100 lower (i.e., the No. 101 was physically identical to the No. 1, but had been graduated with metric divisions on the outside). Apparently the demand for metric rules was not high enough at that time to support their continued listing in the catalogue; all of them were dropped a few years later, after 1879 (Ref. 12).

In 1919, after a hiatus of forty years, Stanley reintroduced metric rules to their standard product line, initially in the form of a 1 Foot metric/inch school rule, the No. 98M. 6 years later, in 1925, two meter sticks, the No. 141 and the No. 142, also with metric/inch graduations, were added. These sticks were the metric equivalents of the No. 41 yard stick, with the fractions of a yard scale replaced with metric graduations. In 1934 these were supplemented by all-metric versions, the Nos. 141M and 142M.

The purpose of these five rules must have been to supply the educational and overseas markets; the English system of measurement was virtually universal in the United States between the two world wars, the metric system only being taught as a curiosity in the public schools (a trend that has persisted almost until the present day).

There does not seem to have been a consistent policy at Stanley regarding left to right vs. right to left graduations on metric rules. The rules offered in 1877–1879 were graduated left to right, as were the No. 98M school rule, and the meter sticks of 1925–1943. A number of examples have been observed, however, of uncatalogued metric rules dating from 1880–1900 which were graduated from right to left, and we can only conclude that the company was indecisive in this area.

By 1941, only three of the five were still in production, and none of these survived the war. All were apparently discontinued to make room for essential production, and when peace returned were not reintroduced.

10 Carpenters' Rule, ½ Meter, 4 Fold

Offered:	1877 thru 1879
Construction:	Arch Joint
Material:	Boxwood, Brass Hinges & Tips
Width:	20 mm.
Graduations:	Centimeters/Millimeters 16ths of Inches

The Carpenters' (Sliding) Rule

The carpenters' sliding rule is a 2 foot, 2 fold rule similar in design to common rules such as the Nos. 1, 2, etc., but with the usual graduations on the back replaced by a brass slide graduated as a simple slide rule and by a set of drafting scales.

The slide rule, in the upper leg, has four logarithmic scales, two (the B and C scales) on the brass slide, and two (the A and D scales) on the wood surface adjacent to it. Three of these scales (the A, B, and C scales) are the same, being two cycles in length, beginning and ending at 1 (that is, nonfolded). The D scale (also labeled GIRT LINE) is only one cycle long, and is folded at 4 (that is, this scale begins at 4 on the left, progresses through 5, 6, etc., to 1 near the middle, and thence through 2, 3, etc., to 4 again on the right.

This is a pattern of slide rule invented about 1677, by Henry Coggeshall in England, and described by him in a pamphlet published in that year (Ref. 101). Coggeshall's arrangement of the scales on the slide rule was particularly adapted for use in timber measurement, allowing easy and rapid computation of the cubic contents of the timber which could be cut from a log. Coggeshall's method was actually a crude form of log rule (see LOG RULES AND LOG MEASURES, later on in this chapter) which estimated the area of the cut timber by using the square of ¼ of the girth of the log (hence the expression GIRT LINE).

The set of drafting scales occupied the entire surface of the lower leg. There were four of them, each 9 inches long, graduated from left to right (that is, beginning near the joint and ending near the tip of the leg), for scale factors of ¼, ½, ¾, and 1 inch per foot. On each scale the first 3 inches were divided into scale feet and inches, and the remaining 6 into scale feet only.

The front face of this rule was graduated in 8ths, 10ths, and 16ths of inches, and was marked with the so-called octagonal or "four-square" lines. Also, if the rule was not bound, the outside edge was graduated in 100ths of a foot.

It is not known when this particular pattern of rule was developed. It must have been after 1677, when Coggeshall invented his slide rule with the Girt Line. It has been suggested tha this rule formed the basis for the engineers' sliding rule, as developed by Routledge, and thus must be earlier than 1811, but this is only a supposition, and needs more research to prove or disprove it. Whatever its origin, the carpenters' (sliding) rule was a highly standardized pattern by the middle of the nineteenth century, produced by a large number of makers, and widely used.

Stanley manufactured four different variations of the carpenters' (sliding) rule, the Nos. 12, 14, 15, and 27. All conformed to the pattern of graduations and Gunter's slide described above; the differences between them were in style of joint and type of trim. Three of the four, the Nos. 12, 14, and 15, had arch joints; the fourth, the No. 27, a square joint. One, the No. 14, was half bound; another, the No. 15, was full bound.

All four of these rules were part of the initial offering of A. Stanley & Co. in 1855 (Ref. 1). The No. 14 was discontinued shortly after the merger with Hall & Knapp (Ref. 3), perhaps due to problems with warping; the others remained in production for another fifty-five years, and were only finally dropped in 1917 (Ref. 32) as a result of the dislocations resulting from the First World War.

12 Carpenters' Sliding Rule, 2 Foot, 2 Fold

Offered: 1855 thru 1915
Construction: Arch Joint, Slide in One Leg Bitted (1860 and after)
Material: Boxwood, Brass Hinge, Slide, & Trim
Width: 1½ Inch
Graduations: Body: 8ths, 10ths, and 16ths of Inches
100ths of a Foot (On Outside Edge)
Slide: 8ths of Inches (On Back)
Scales: Body: Gunter's, 2 Cycles
Gunter's, 1 Cycle, Folded at 4
Drafting
Octagonal
Slide: Gunter's, 2 Cycles

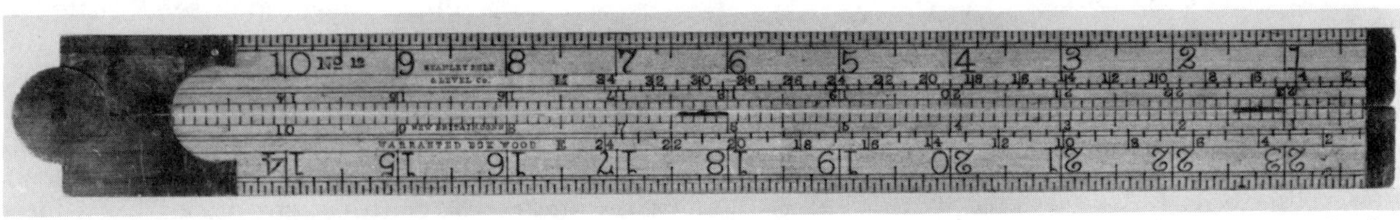

Figure 62a: No. 12, 1858–1915 (Front) *(Charles Leverone Collection)*

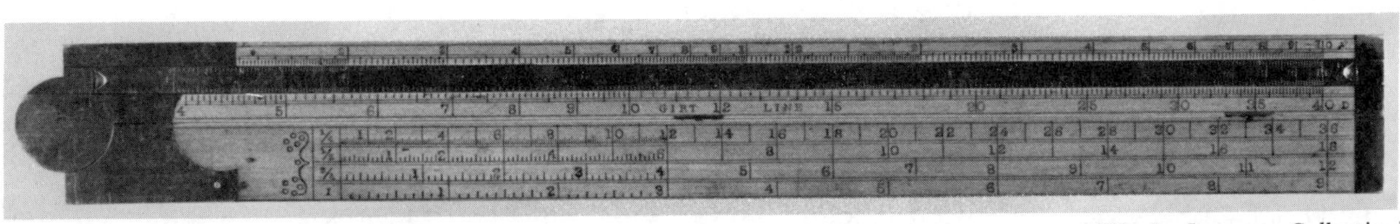

Figure 62b: No. 12, 1858–1915 (Back) *(Charles Leverone Collection)*

13 Carpenters' Caliper Rule, 6 Inch, 2 Fold

Offered: 1884 thru 1932
Construction: Square Joint, Left Hand Caliper
Bitted (1912 thru 1922)
Material: Boxwood, Brass Hinge, Caliper, & Trim
Width: 1⅛ Inch
Graduations: Body: 8ths and 16ths of Inches
Caliper: 16ths of Inches (1922 and before)
16ths and 32nds of Inches (1925 and after)

The No. 13 was one of the four rules which were retained when Stanley discontinued the rest of the Stearns line of boxwood rules in 1884 (Ref. 13); the others were the Nos. 30, 31, and 80. Stanley had not hitherto offered an extra wide 6 inch caliper rule, and the catalogue description of this "new" rule (except for the one difference noted below) corresponded exactly with that of the Stearns No. 13, last listed in 1879 (Ref. 12). It is highly probable that this was not a coincidence, but simply represented the transfer to the Stanley product line of a rule too popular to be eliminated with the others.

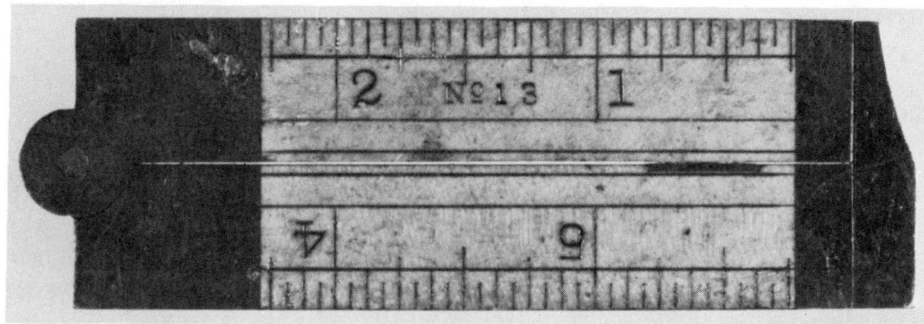

Figure 63a: No. 13, 1912–1921

(Roger K. Smith Collection)

The one physical difference between the Stearns No. 13 and its Stanley equivalent was in the shape of the back of the caliper jaw; where the Stearns caliper jaw had a straight taper on the back (see Fig. 63b), that on the Stanley No. 13 had a hollow taper.

This change was probably made for the sake of uniformity, so that the appearance of the No. 13 would correspond to that of the other wide caliper rules in the product line, all of which already had a hollow curve, and to reduce manufacturing cost by using a standard shape jaw.

After its inclusion in the Stanley line, the No. 13 was produced for nearly fifty years, indicating that it must have been a popular rule (probably because of its greater than usual depth of throat). It was finally discontinued between 1932 (Ref. 43) and 1934 (Ref. 44), a period when many rules were dropped as a result of the depression of 1929–1939.

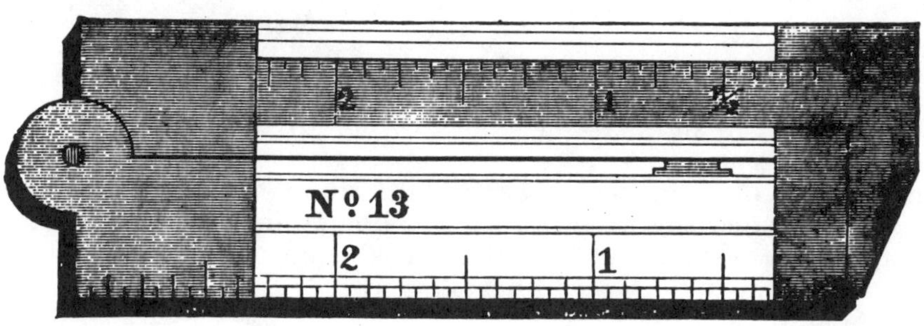

Figure 63b: Stearns No. 13 Caliper Rule, 1867

(Ref. 7)

13½ Carpenters' Caliper Rule, 6 Inch, 2 Fold

Offered: 1888 thru 1941

Construction: Square Joint
Left Hand Caliper (1932 and before)
Right Hand Caliper (1934 and after)
Bitted (1912 thru 1922)

Material: Boxwood, Brass Hinge, Caliper, & Trim

Width: 1½ Inch

Graduations: Body: 8ths and 16ths of Inches
Caliper: 16ths of Inches (1922 and before)
16ths and 32nds of Inches (1925 and after)

The No. 13½ caliper rule was, at 1½ inches, the widest of the Stanley folding caliper rules. This extra width had its advantages; the wider the rule, the longer the caliper jaw, and thus the larger diameter circular cross-section which could be measured. With its extra width, the No. 13½ could measure circles up to 2 inches in diameter. This was exceptional measuring capacity in a 6 inch, 2 fold rule which, when folded, could be carried in a vest pocket.

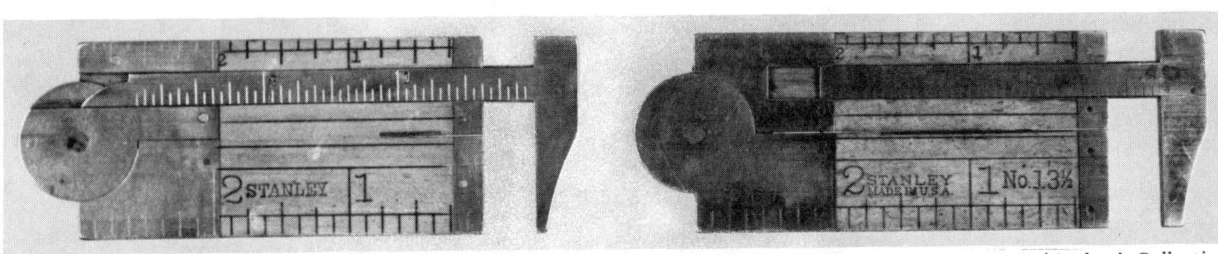

(Author's Collection)

Figure 64a: Nos. 13½, 1919 and Before (Left), 1922 and After (Right)

This combination of compactness and measuring capacity created special problems in construction, however. The normal configuration for a 2 fold caliper rule was to have the rule joint be a square joint, and have the caliper slide extend from the end of the leg to the joint plates (see Figure 64b).

(Author's Collection)

Figure 64b: Typical 2 Fold Caliper Rule Construction (No. 36½L)

If the No. 13½ had been made in this way, the joint plates, extending up the leg for more than 1 inch, would have caused the length of the caliper slide to be less than 2 inches. When using the caliper some of the slide must remain in the leg of the rule; a slide of this length would have limited the range of the caliper to 1½ inches or less, significantly restricting the range otherwise made possible by the 1 inch caliper jaw.

For these reasons, the No. 13½ was fitted with an atypical rule joint which allowed the caliper slide to be at least 2½ inches in length.

Initially, the rule joint used was the same as that on the 2 foot, 2 fold rules with slide (No. 26 *et al.*), with the slide extending fully through the joint plate and shaped on the end to conform to the profile of the curved/straight end of the leg (Figure 64a [Left]).

When the other rules using this joint were discontinued after 1915, the joint on the No. 13½, no longer constrained by the need for manufacturing conformity to be compatible with any other rule, was modified. The new joint, introduced in 1922, was designed so that the caliper slide only extended half way through the joint plate and was square on the end (Figure 64a [Right]). This compromise significantly strengthened the joint while still leaving enough length on the caliper slide, and was the form in which the rule was manufactured from then on.

In 1934, the No. 13½ was switched from having a left hand caliper (see SLIDES AND CALIPERS, Chapter 4) to having a right hand caliper (Figure 64c). This change was made at this time to all Stanley caliper rules still in production, apparently in response to market demand.

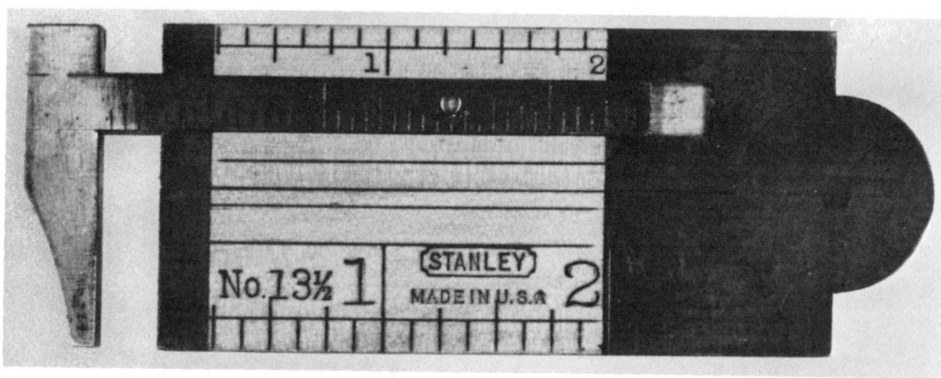

Figure 64c: No. 13½, 1934-1941

(William Baader Collection)

The No. 13½ was discontinued after 1941, probably as a result of the rising demand for military production, and was not reintroduced, as many rules were, after the end of hostitlities in 1947.

14 Carpenters' Sliding Rule, 2 Foot, 2 Fold

Offered:	1855 thru 1859	
Construction:	Arch Joint, Half Bound, Slide in One Leg	
Material:	Boxwood, Brass Hinge, Slide, & Trim	
Width:	Not Known, but probably 1½ Inch	
Graduations:	Body:	8ths, 10ths, and 16ths of Inches
	Slide:	8ths of Inches (On Back)
Scales:	Body:	Gunter's, 2 Cycles
		Gunter's, 1 Cycle, Folded at 4
		Drafting
		Octagonal
	Slide:	Gunter's, 2 Cycles

The No. 14 was never illustrated in any catalogue, and no example is known to exist; hence the available information is only the fragmentary facts which can be elicited from the written description. The data given here as regards the scales and features of this rule is based largely on the assumption that it was a carpenters' sliding rule of the same pattern as the others made by Stanley. This is probably a valid assumption; the other three rules with slides (discounting the No. 26, which was described as having a "Plain Slide") were all carpenters' sliding rules, and it is highly likely that the No. 14 was also.

Half bound 2 foot, 2 fold rules were only part of the Stanley product line for a very short time. Only two were made, the No. 3 (without slide) and the No. 14 (with slide), both offered from 1855 (Ref. 1) through 1859 (Ref. 2).

It is not known why Stanley discontinued these two rules after only a short period. Other half bound rules, of the 2 foot, 4 fold type, were apparantly well received, and remained in the product line for many years (more than 70 years in the case of the No. 84).

It is possible that these two rules were discontinued due to problems with warping. Binding only one edge of a piece of boxwood with brass could cause unbalanced stress as the wood expands and contracts with varying humidity, and the stresses would be more severe on a 1 foot piece than on one only half as long.

14 Carpenters' Caliper Rule, 6 Inch, 2 Fold

Offered:	1909 thru 1917	
Construction:	Square Joint, Full Bound, Left Hand Caliper	
Material:	Boxwood, Brass Hinge, Caliper, & Trim	
Width:	7/8 Inch	
Graduations:	Body:	8ths, 10ths, 12ths, and 16ths of Inches
	Caliper:	16ths of Inches

15 Carpenters' Sliding Rule, 2 Foot, 2 Fold

Offered:	1855 thru 1915	
Construction:	Arch Joint, Full Bound, Slide in One Leg	
Material:	Boxwood, Brass Hinge, Slide, & Trim	
Width:	1½ Inch	
Graduations:	Body:	8ths, 10ths, and 16ths of Inches
		100ths of a Foot (1909)
	Slide:	8ths of Inches (On Back)
Scales:	Body:	Gunter's, 2 Cycles
		Gunter's, 1 Cycle, Folded at 4
		Drafting
		Octagonal
	Slide:	Gunter's, 2 Cycles

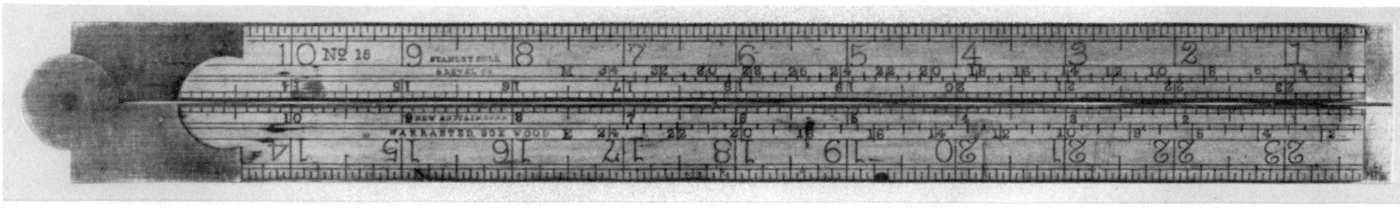

Figure 65a: No. 15, 1858–1915 (Front) *(Author's Collection)*

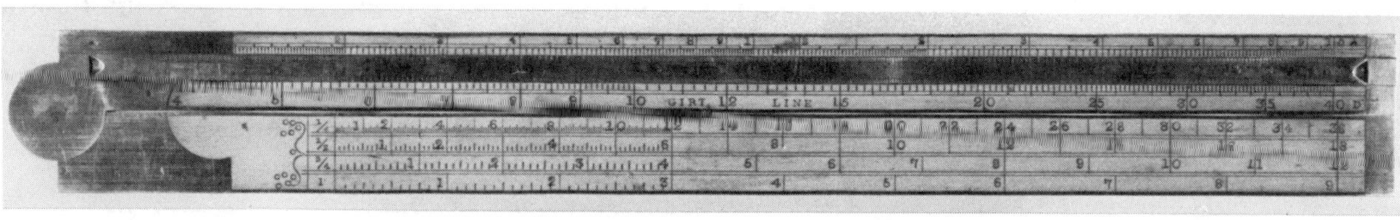

Figure 65b: No. 15, 1858–1915 (Back) *(Author's Collection)*

16 Engineers' Sliding Rule, 2 Foot, 2 Fold

Offered: 1855 thru 1902
Construction: Arch Joint, Full Bound, Slide in One Leg
Material: Boxwood, Brass Hinge, Slide, & Trim
Width: 1½ Inch
Graduations: Body: 8ths, 10ths, and 16ths of Inches
Slide: 8ths of Inches (On Back)
Scales: Body: Gunter's, 2 Cycles
Gunter's, 1 Cycle
Octagonal
Slide: Gunter's, 2 Cycles
Other Markings: Volume, Weight, Polygon, Circle, & Pumping Engine Tables

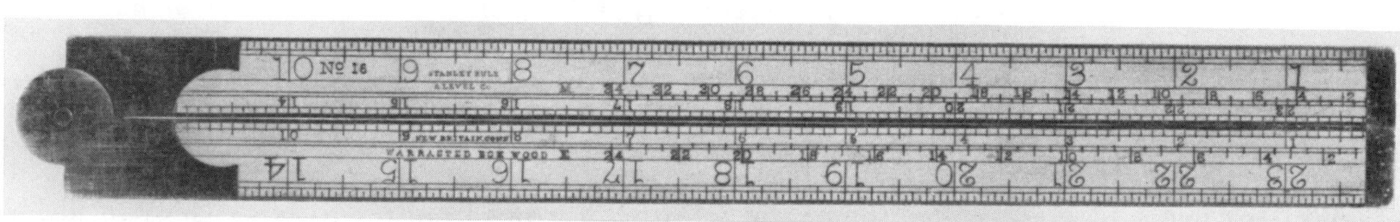

Figure 66a: No. 16, 1858–1902 (Front) *(Author's Collection)*

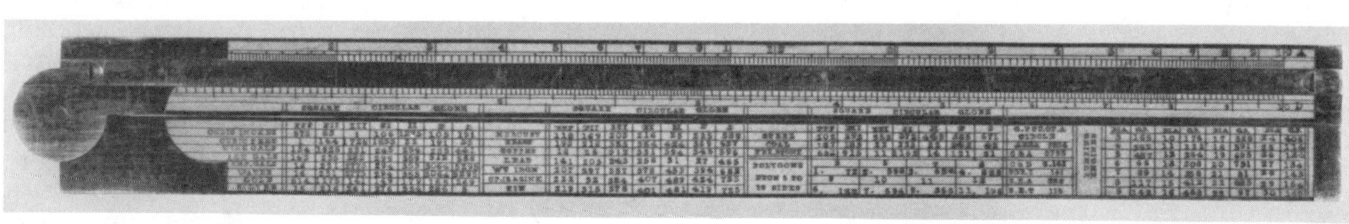

Figure 66b: No. 16, 1858–1902 (Back) *(Author's Collection)*

17 Blacksmiths' Rule, 2 Foot, 2 Fold

Offered: 1900 thru 1941
Construction: Concealed Joint (1917 and before)
Rivet Joint (1919 and after)
Material: Spring Brass, Steel Joint
Width: 5/8 Inch
Graduations: 8ths and 16ths of Inches

The blacksmiths' 2 Foot, 2 Fold rule was traditionally made of brass. It could not practically be made of wood; in the process of making measurements on hot iron a wood rule would quickly be scorched or set aflame. Steel would have resisted the heat, but would have to be frequently cooled in the blacksmiths' water barrel, and very soon be too rusty to read. Brass, on the other hand, could easily be kept bright with an occasional light polishing.

(Ref. 32)

Figure 67a: No. 17, 1917

The No. 17 is a classic example of improving a product just as the demand for it begins a drastic decline. In this case the improvement was the new spring joint, and the declining demand the result of the shrinkage in the demand for blacksmiths which had begun twenty years earlier, and would end within twenty more years with the virtual elimination of this trade. As Aldren Watson points out in his book, *The Village Blacksmith* (Ref. 132):

By the turn of the century, blacksmithing had begun a sharp descent on the incline to virtual extinction. The gasoline engine and the proliferation of new developments in the iron and steel industry were creating changes of such a radical nature that in the brief period from 1900 to 1930 innovations in manufacturing, transportation, and communication would make nearly obsolete not only the blacksmith and his work but also the kinds of articles that had been in demand during the simpler life of the nineteenth century.

The steel spring joint used in the No. 17 was a significant departure from the rule and knuckle joints used on ordinary folding rules, and from the friction joint previously used for blacksmiths' rules. A German invention, this spring joint had recently been licensed in the United States by Stanley for use in their line of "Zig-Zag" wooden rules introduced at this same time, the type of rule which would come to dominate the United States rule market in the next thirty years.

Two versions of this joint were used in the blacksmiths' rule. The concealed joint, in which there is no hole through the brass, the rivet only penetrating the steel joint plates and spring, was used first. Later, between 1917 and 1919, this was replaced by the rivet joint, in which the rivet was carried through both the joint plates and spring, and the legs as well.

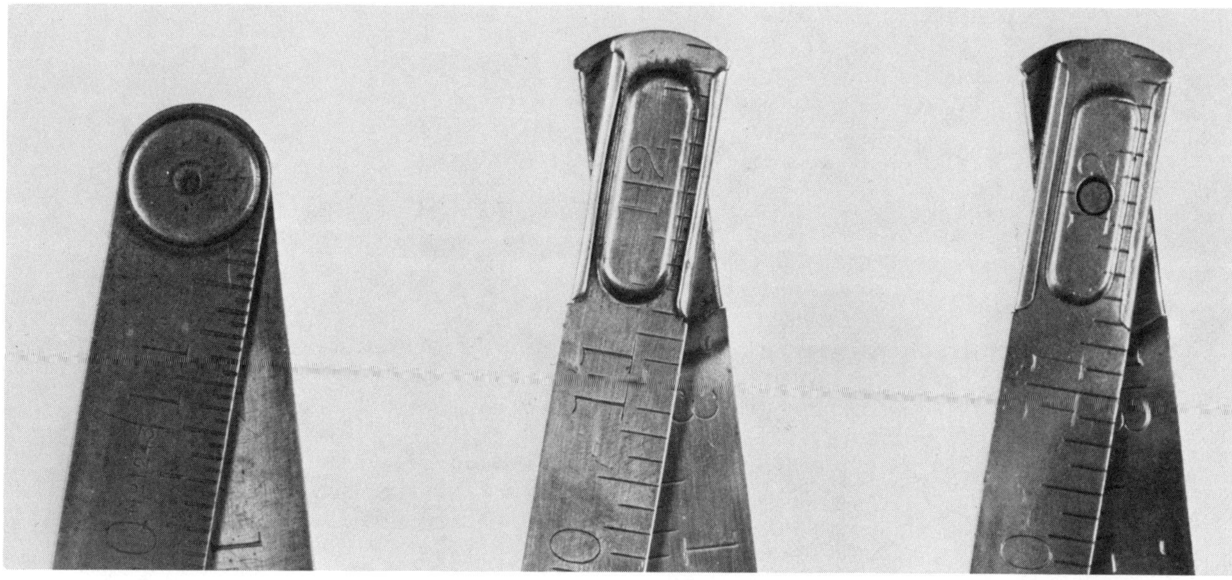

Figure 67b: Comparison of Friction Joint (Left), Concealed Spring Joint (Center), and Rivet Spring Joint (Right) *(Author's Collection)*

18 Carpenters' Rule, 2 Foot, 2 Fold

Offered: 1855 thru 1949
Construction: Square Joint
Material: Boxwood, Brass Hinge & Tips
Width: 1½ Inch (1943 and before)
1⅜ Inch (1947 and after)
Graduations: 8ths and 16ths of Inches

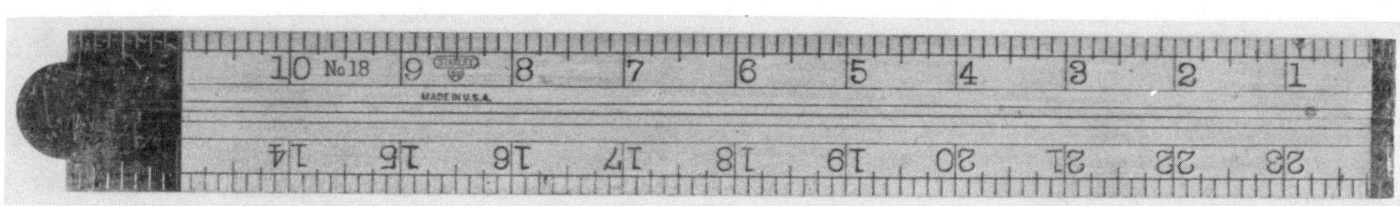

Figure 68: No. 18, 1922–1932 *(Author's Collection)*

19 Carpenters' Rule, 2 Foot, 2 Fold

Offered: 1855 thru 1859
Construction: Square Joint, Bitted
Material: Boxwood, Brass Hinge & Trim
Width: Not Known, but probably 1½ Inch
Graduations: Not Known

20 Carpenters' Rule, 1 Meter, 4 Fold

Offered:	1877 thru 1879
Construction:	Arch Joint
Material:	Boxwood, Brass Hinges & Tips
Width:	25 mm.
Graduations:	Centimeters/Millimeters 16ths of Inches

22 Carpenters' Rule, 2 Foot, 2 Fold

Offered:	1855 thru 1915
Construction:	Square Joint Bitted (1860 and after)
Material:	Boxwood, Brass Hinge & Trim
Width:	1½ Inch
Graduations:	16ths of Inches (1872 and before) 10ths and 16ths of Inches (1874 thru 1877) 10ths, 12ths, and 16ths of Inches (1879 and after)
Scales:	Octagonal
Other Markings:	Board Measure Tables

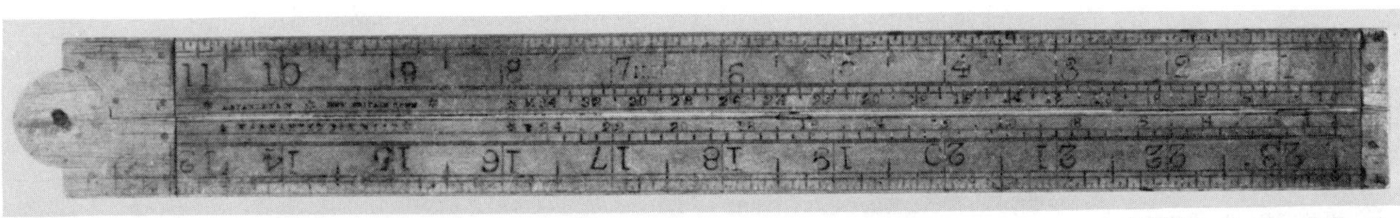

Figure 69a: No. 22, 1854–1857 (Front) *(William Baader Collection)*

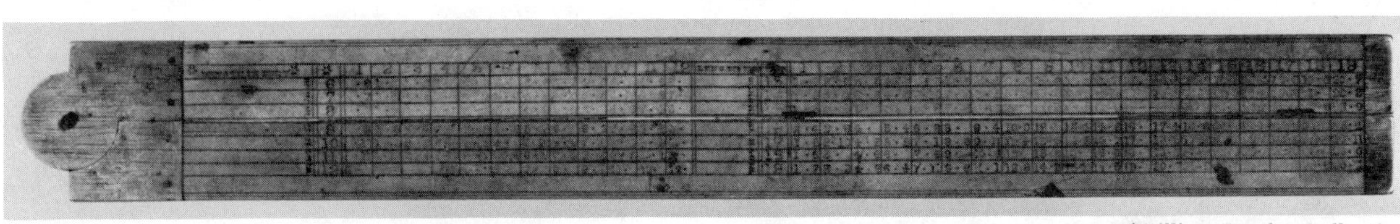

Figure 69b: No. 22, 1854–1857 (Back) *(William Baader Collection)*

The Button Gauge

The Button Gauge is a short caliper rule used by tailors and seamstresses to measure buttons so as to be able to size the associated buttonhole. Typically about three inches in length, these devices were provided with a sliding caliper calibrated in 40ths of an inch, the commonly accepted scale for button size.

Stanley offered button gauges twice during the company's early years: the No. 210, in 1862 (Ref. 5), and the Nos. 23 and 24 in 1867 (Ref. 7). The Nos. 210 and 23 were made of boxwood, the No. 24, of ivory. All had brass calipers and trim. It is probable that the Nos. 210 and 23 were identical, the second rule being simply the first renumbered after a short absence from the product line.

Button gauges were not offered after 1867. This was a period when Stanley was eliminating extraneous products in order to focus more specifically on tools, and apparently it was felt that a button gauge was too far afield.

23 Caliper Button Gauge, 4 Inch

Offered:	1867
Construction:	Right Hand Caliper
Material:	Boxwood, Brass Caliper & Tips
Width:	7/8 Inch
Graduations:	Body: Not Known
	Caliper: 40ths of Inches

Figure 70: No. 23, 1867 *(William Baader Collection)*

The School Rule

The school rule is a short, nonfolding rule intended for desk use, either as a straightedge, or for simple measuring operations. Also referred to as a "scholars'," or "desk" rule, it is almost invariably 1 foot in length, to allow easy carrying in a notebook or satchel. Over the years such rules have been produced in a wide selection of woods, graduations, widths, and construction details by a number of firms, one of which was the Stanley Rule & Level Co.

A. Stanley offered one such rule in 1855 (Ref. 1), the No. 50. It was described in the price list as a boxwood "scholars'" rule, but little else is known about it as to width, graduations, or construction. This rule was offered until 1860 (Ref. 3), after which it was dropped.

Two years later, in 1862 (Ref. 5), Stanley began to offer a line of stationers' goods, including a whole line of rulers, the Nos. 160, 162, 164, 166, 168, and 170. The first five of these, simply described as "rulers," were either flat, round, or octagonal, and were made of either mahogany or rosewood. These were probably ungraduated accountants' rulers, used to rule the lines in account books. The No. 170, however, was flat and made of boxwood, and was called a "scholars'" rule like the No. 50. This line of rules was also short-lived, however, and was discontinued after 1865 (Ref. 6).

Three more school rules were offered in the next few years, the Nos. 98 and 99 "school" rules beginning in 1867 (Ref. 7), and the No. 23 "scholars'" rule in 1874 (Ref. 10). These three rules were identical as to construction, being ¾ inch wide with a beveled edge, but differed as to material and graduations. The No. 23 was made of maple, and graduated in 16ths of inches; the No. 98 of boxwood, in 8ths and 16ths; and the No. 99, also of boxwood, in 10ths and 16ths.

All three of these rules were discontinued after 1877, perhaps due to reasons of cost. School rules were not a premium product, but were intended to be cheaply made and inexpensive. After trying boxwood rules in this market in 1867, and less expensive maple ones seven years later, Stanley probably concluded that there was no room in this market for a quality product, and abandoned it to other manufacturers with lower standards.

In 1919, more than forty years later, a second attempt was made to manufacture and market school rules.

Four rules were offered. Two, the Nos. 98 and 98M "desk" rules, were revivals of the No. 98 rule discontinued in 1877, differing only in their graduations; the No. 98 was graduated in 8ths and 16ths, the No. 98M in 8ths and millimeters.

The other two rules were new. These were the Nos. 34¼ and 34½ "school" rules. Differing in material (The No. 34¼ was made of maple, the No. 34½ of boxwood), they were otherwise alike, both being flat, 1⅛ inches wide, with brass tips.

As in 1867–1877, these four rules must have been only marginally profitable items in the product line. In 1934, when the economic pressure of the depression was being fully experienced, all further efforts to market school rules were ended. The Nos. 98 and 98M were dropped from the line; the Nos. 34¼ and 34½ were changed to have vertical figures, and sold for carpenters' use as the Nos. 34¼V and 34½V "bench" rules. In this market, both were reasonably successful; the No. 34½V was manufactured until 1943 (Ref. 53), when supplies of boxwood became scarce, and the No. 34¼V was produced well into the 1970's (Ref. 68).

23 School Rule, 1 Foot

Offered:	1874 thru 1877
Construction:	Beveled Edge
Material:	Maple
Width:	¾ Inch
Graduations:	16ths of Inches

24 Caliper Button Gauge, 4 Inch

Offered:	1867	
Construction:	Right Hand Caliper	
Material:	Ivory, Brass Caliper & Tips	
Width:	⅞ Inch	
Graduations:	Body:	Not Known
	Caliper:	40ths of Inches

The No. 24 was never illustrated in any catalogue, and no example is known to exist; hence the available information is only the fragmentary facts which can be elicited from the written description. The data given here as regards its scales and dimensions is based on the assumption that, except for being made of ivory, it was identical to the No. 23. This is probably a reasonable surmise; the two gauges were introduced at the same time, and had the same short period of availability, and were probably alike in the same way as the Nos. 36 and 38.

26 Carpenters' Rule With Extension Slide, 2 Foot, 2 Fold

Offered:	1855 thru 1917	
Construction:	Square Joint, Slide in One Leg	
Material:	Boxwood, Brass Hinge, Slide, & Trim	
Width:	1½ Inch	
Graduations:	Body:	8ths, 10ths, and 16ths of Inches
	Slide:	8ths of Inches
Scales:	Body:	Octagonal

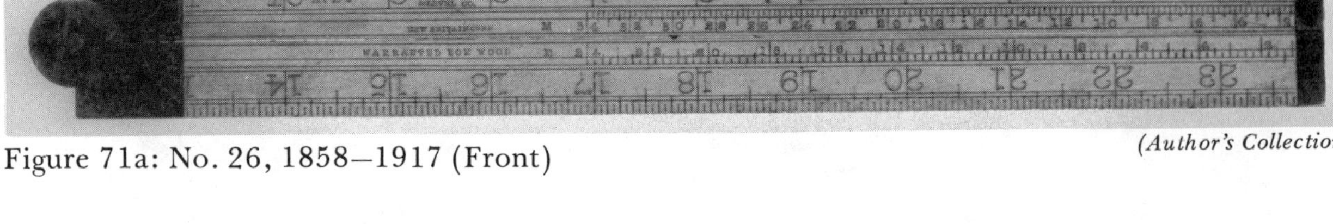

Figure 71a: No. 26, 1858–1917 (Front) *(Author's Collection)*

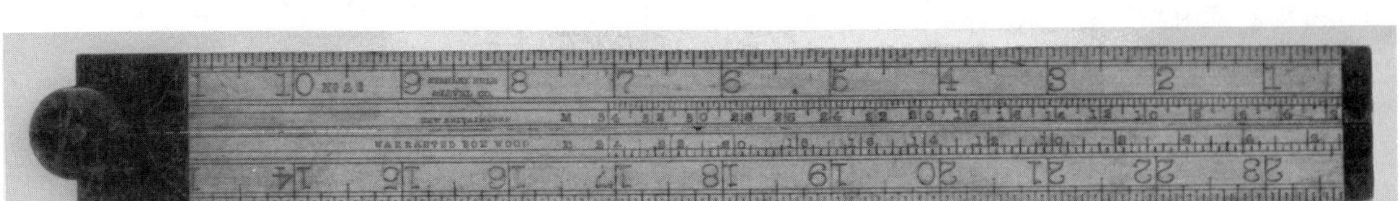

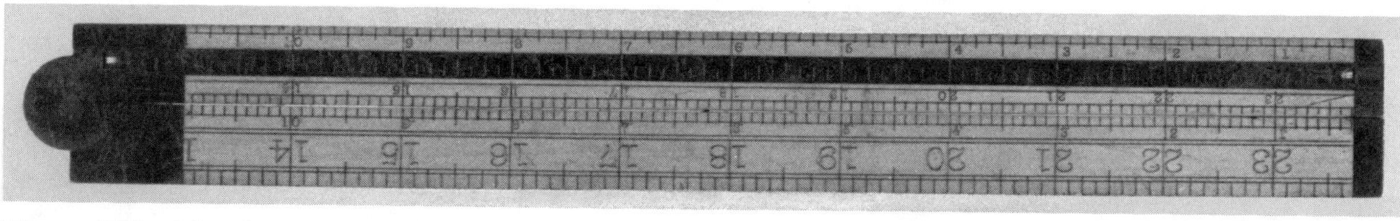

Figure 71b: No. 26, 1858–1917 (Back) *(Author's Collection)*

27 Carpenters' Sliding Rule, 2 Foot, 2 Fold

Offered:	1855 thru 1915	
Construction:	Square Joint, Slide in One Leg	
	Bitted (1860 and after)	
Material:	Boxwood, Brass Hinge, Slide, & Trim	
Width:	1½ Inch	
Graduations:	Body:	8ths, 10ths, and 16ths of Inches
		100ths of a Foot (On Outside Edge)
	Slide:	8ths of Inches (On Back)
Scales:	Body:	Gunter's, 2 Cycles
		Gunter's, 1 Cycle, Folded at 4
		Drafting
		Octagonal
	Slide:	Gunter's, 2 Cycles

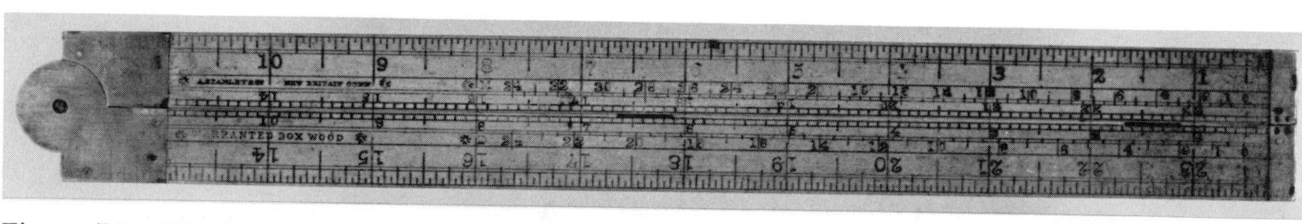

Figure 72a: No. 27, 1854–1857 (Front) *(William Baader Collection)*

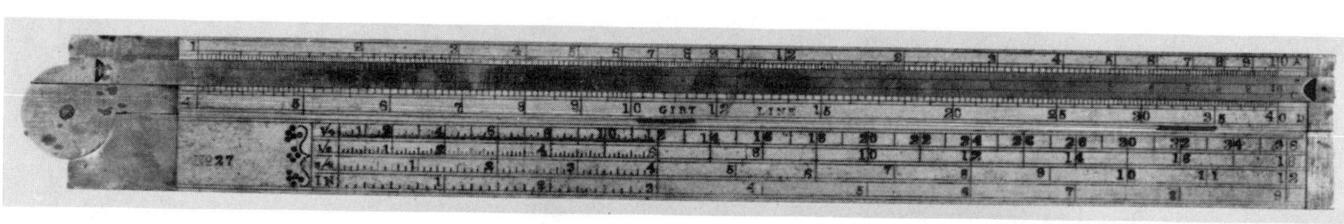

Figure 72b: No. 27, 1854–1857 (Back) *(William Baader Collection)*

27 Carpenters' Rule, 2 Foot, 4 Fold

Offered:	1932 thru 1941, 1947 thru 1950
Construction:	Round Joint, Middle Plates Vertical Figures (1947 and after)
Material:	Maple, Brass Hinges, Steel Tips
Width:	1 Inch
Graduations:	8ths and 16ths of Inches

In 1942 the No. 27, along with many other rules, was discontinued to conserve materials and to make room for more important war production. The price list (Ref. 52) described this cancellation as only "temporary," and although some other rules so canceled were never returned to production, this was not the case with the No. 27, and it was back in the product line when the first postwar catalogue was published in 1947 (Ref. 54). This reprieve was only temporary, however, and within three years this rule was canceled again, this time permanently, in the product line retrenchment of the 1950's.

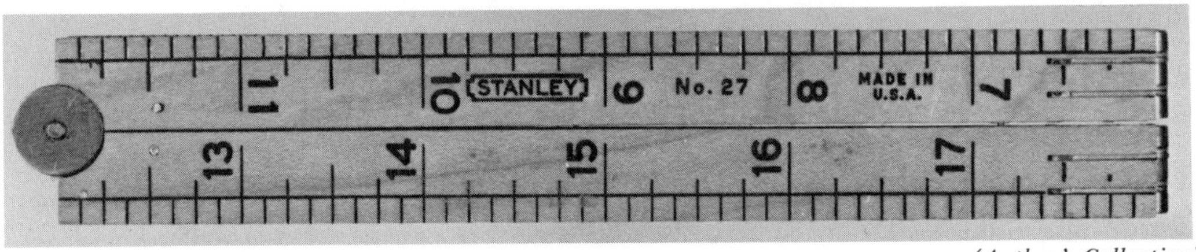

(Author's Collection)

Figure 73: No. 27, 1947–1950

28 Carpenters' Rule, 2 Foot, 2 Fold

Offered:	1855 thru 1859
Construction:	Square Joint, Full Bound
Material:	Boxwood, Brass Hinge & Trim
Width:	Not Known, but probably 1½ Inch
Graduations:	Not Known

29 Carpenters' Rule, 2 Foot, 2 Fold

Offered:	1855 thru 1915
Construction:	Round Joint
Material:	Boxwood, Brass Hinge & Tips
Width:	1³/₈ Inch
Graduations:	8ths and 16ths of Inches

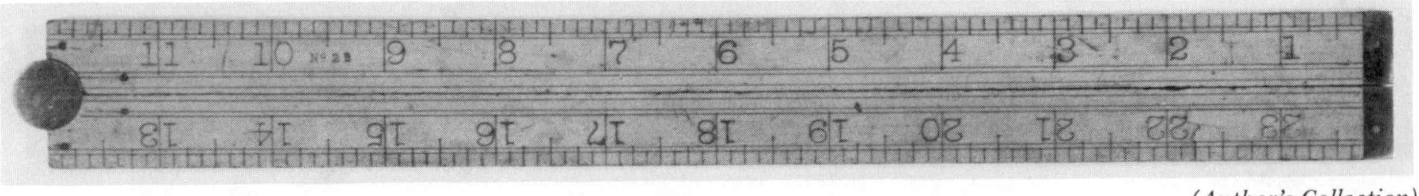

(Author's Collection)

Figure 74: No. 29, 1858–1915

30 Carpenters' Rule, 6 Inch, 2 Fold

Offered:	1859
Construction:	Round Joint
Material:	Boxwood, Brass Hinge & Tips
Width:	Not Known
Graduations:	Not Known

30 Carpenters' Rule, 1 Meter, 4 Fold

Offered:	1877 thru 1879
Construction:	Arch Joint, Middle Plates
Material:	Boxwood, Brass Hinges & Tips
Width:	34 mm.
Graduations:	Centimeters/Millimeters 16ths of Inches

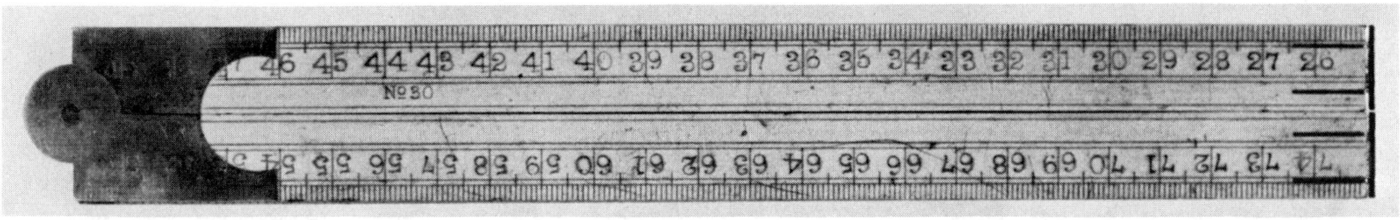

Figure 75: No. 30, 1877–1879 *(Author's Collection)*

Shrinkage Rules

The shrinkage (or contraction) rule is an important item in the set of specialized tools used by the patternmaker in the production of patterns for foundry use.

Such patterns, hitherto made primarily of wood, but more recently also of metal and plastic, are models of objects which are intended to be cast from molten metal. The pattern is used to form a hollow mold by packing damp sand around it, and then withdrawing it. The hot liquid metal is poured into the resultant cavity, where it cools and solidifies into a copy of the pattern.

As the metal in the mold solidifies and cools, it also shrinks, with the result that the copy is measurably smaller than the pattern. To compensate for this shrinkage, the pattern is deliberately made oversize by a percentage appropriate to the melting point and expansion coefficient of the particular alloy being cast. This shrinkage allowance can be as little as 1/16 or as great as 11/16 inch per foot.

The problem of shrinkage allowance is further complicated by the fact that there is frequently an intermediate stage in the pattern to casting sequence. It is often the case that the original pattern will be used to cast a more durable metal pattern of, say, iron or aluminum, and that this secondary pattern will in turn be used to make the final casting. In such a case, two shrinkage allowances, added together, would have to be used in making the original pattern (e.g., if the intermediate pattern was iron (1/8 inch per foot), and the final casting brass (3/16 inch per foot), then the original pattern would have to be made with a shrinkage allowance of 5/16 inch per foot.

Typical shrinkage allowances for different alloys and alloy combinations are as shown in the following table:

Alloy or Alloy Combination	Shrinkage Allowance (per Foot)
Aluminum	3/16 Inch
Aluminum/Brass	3/8 Inch
Bismuth	5/32 Inch
Brass	3/16 Inch
Britannia Metal	1/32 Inch
Cast Iron	1/8 Inch
Cast Iron/Brass	5/16 Inch
Cast Iron/Cast Iron	1/4 Inch
Copper	3/16 Inch
Lead	5/16 Inch
Malleable Iron	1/8 Inch
Steel	1/4 Inch
Tin	1/12 Inch
Zinc	5/16 Inch

Figure 76a: Table of Typical Shrinkage Allowances

To compute and apply this shrinkage allowance to all dimensions during the construction of the pattern would be a tedious and error prone task. Instead, the patternmaker uses a shrinkage rule (a rule whose graduations are slightly oversize to include the shrinkage allowance) to dimension the pattern. Such rules are graduated as if they were their nominal length, but are physically longer by the shrinkage allowance. For example, a 2 foot shrinkage rule for cast iron would be graduated into 24 "inches," but would actually be 24¼ inches in length.

Shrinkage rules were identified in different ways. In the United States the most common method was to mark them with the shrinkage allowance in fractions of an inch per foot (the method used in the table above). An alternative method less commonly used was to mark them with a figure indicating the shrinkage allowance relative to the shrinkage allowance for cast iron (i.e., 1 "shrink" = 1/8 inch per foot, 1½ "shrinks"= 3/16 inch per foot, 2 "shrinks"= 1/4 inch per foot, etc.). The continental custom was to mark the rule with a proper fraction representing the shrinkage allowance in "feet per foot" (or "meters per meter," etc.) (i.e., 1/96 = 1/8 inch per foot, 1/64 = 3/16 inch per foot, 1/48 = 1/4 inch per foot, etc.).

A patternmaker would have a whole set of such rules, each with a different shrinkage allowance, to allow for all possible requirements imposed by different jobs.

Figure 76b: Shrinkage Rules With Different Allowances *(Author's Collection)*

Prior to the middle of the nineteenth century, patternmakers would make their own shrinkage rules, beginning with common contraction factors such as those for iron and brass (a method for doing this by scaling from a standard rule is given in Chase, Chapter XXXV [Ref. 114]). Even after commercially made shrinkage rules became available, this was still done if the required shrinkage allowance was non-standard.

Typically, a shrinkage rule would be made of boxwood, (nominally) 2 feet long, and about 1½ inches wide. They were made both nonfolding, and in a 2 fold form, and were invariably graduated from left to right along the top of each of the four edges. The most common graduations were 8ths and 16ths of inches, but 10ths and 12ths were also used, particularly after about 1890.

Between 1850 and 1910, shrinkage rules were made in an increasing number of contraction allowances by Stanley, and other companies such as D.H. Stephens & Co., the Lufkin Rule Co., and the Keuffel & Esser Co. As a result, the practise of making one's own rules became less common, and most patternakers' tool kits from post-1900 will contain few, if any, craftsman-made shrinkage rules.

In the period after World War I, pattern making became more and more a precision trade, assuming aspects similar to those of machinist or diemaker, and the precision possible using wooden measuring instruments graduated to 16ths of an inch was no longer adequate. As a result, during this period the wood shrinkage rule was supplanted by steel shrinkage rules produced by the manufacturers of machinists' tools, such as the L.S. Starrett Co., and the Brown & Sharpe Co.

The last wood shrinkage rules were manufactured in the 1940's.

30 Patternakers' Shrinkage Rule, 24¼ Inch

Offered: 1884 thru 1932
Material: Boxwood, Brass Tips
Width: 1¼ Inch
Graduations: 8ths and 16ths of Inches @ 1/8 Inch per Foot Shrinkage

The No. 30 was another of the four rules which were retained when Stanley discontinued the rest of the Stearns line of wood rules in 1884 (Ref. 13) (the others were the Nos. 13, 31, and 80). Stanley had not offered shrinkage rules prior to that time, and the catalogue description of this "new" rule corresponded exactly with that of the Stearns No. 81, which was listed for the last time in 1879 (Ref. 12). It is highly probable that this was not a coincidence, but was simply the transfer to the Stanley product line of a rule too popular to be eliminated along with the Stearns name.

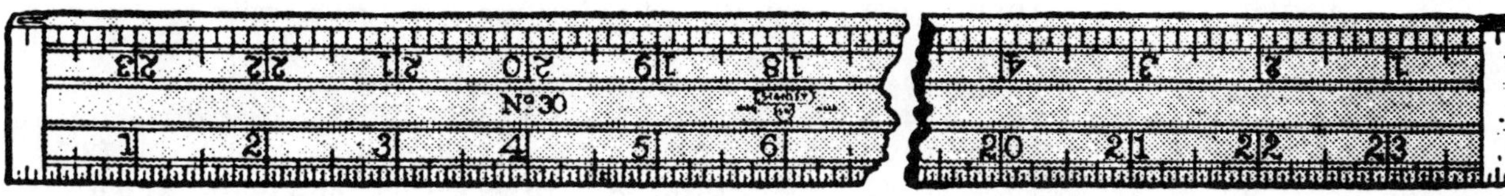

Figure 77: No. 30, 1926 *(Ref. 39)*

Initially, and up until 1892 (Ref. 15), the No. 30 was only offered inone shrinkage allowance, 1/8 inch per foot (the allowance for cast iron). In 1898 (Ref. 17), a number of other shrinkage allowances (1/16 , 1/10 , 3/32 , 3/16 , and 1/4 inch per foot) were also made available by specail order. In 1900, when the Nos. 30½ and 31½ shrinkage rules were introduced in a wide range of shrinkage allowances, this offer of special order graduations for the No. 30 was withdrawn (Ref. 18). The No. 30 continued to be offered until 1932 (Ref. 43), but after 1900, only with its original 1/8 inch per foot shrinkage allowance.

30½A Patternmakers' Shrinkage Rule, 24½ Inch

Offered: 1900 thru 1932
Material: Boxwood, Brass Tips
Width: 1½ Inch
Graduations: 8ths, 10ths, 12ths, and 16ths of Inches @ 1/16 Inch per Foot Shrinkage

The primary differences between the No. 30½ shrinkage rule and the No. 30, which had been in production for many years, were two: the No. 30½ was wider (1½ vs. 1¼ inches), and thus presumably stronger, and it was graduated in 8ths, 10ths, 12ths, and 16ths of inches, instead of only 8ths and 16ths.

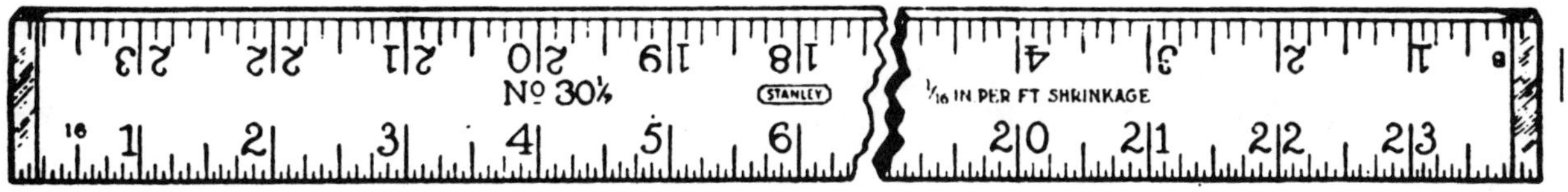

Figure 78a: No. 30½A, 1938 *(Ref. 48)*

As first manufactured, the No. 30½ had the same elaborate longitudinal embellishment lines as the No. 30 (see Fig. 80). These were discontinued within two years, however, and from 1902 on the surface of the rule was left plain except for the graduations.

Over its lifetime the No. 30½ was produced in a dozen different shrinkage allowances. Initially seven were available; later this number was expanded to twelve; and still later reduced, first to ten, and then to six. From 1909 on (Ref. 25) the shrinkage allowance was distinguished by a letter suffix (A through M, omitting I) attached to the rule number (hence 30½A). The shrinkage allowances offered, together with their associated letter suffixes and the years they were available, are shown in the following table:

Shrinkage Allowance	Letter Suffix	Years Offered
1/16 Inch/Foot	A	1900–1932
1/12 Inch/Foot	B	1900–1932
3/32 Inch/Foot	D	1900–1922
1/10 Inch/Foot	C	1900–1943
1/8 Inch/Foot	E	1900–1943
5/32 Inch/Foot	H	1912–1932
3/16 Inch/Foot	F	1900–1943
1/5 Inch/Foot	J	1912–1922
7/32 Inch/Foot	K	1912–1932
1/4 Inch/Foot	G	1900–1943
5/16 Inch/Foot	L	1912–1943
3/8 Inch/Foot	M	1912–1943

Figure 78b: Shrinkage Allowances of No. 30½ Rules

At first the letter suffix was only used in the catalogue to facilitate ordering, the rules themselves only being marked with the rule No. (30½) and the shrinkage allowance (see Fig. 79). Only later, some time before 1917 (Ref. 32), was the letter added to the number on the rule itself (see Fig. 82).

Until 1932 (Ref. 43) the No. 30½ was available graduated in 8ths and 16ths only at no extra cost. When so ordered, both scales on one side were in 8ths, and both on the other in 16ths. This option must have been fairly popular; at least 20 percent of the No. 30½ rules known to the author are graduated in this manner. Perhaps patternmakers found the 10ths and 12ths scales too easy to confuse, and not frequently required, and chose to omit them.

It is interesting to note that there apparently was no standard as to which side (8ths/16ths or 10ths/12ths), and with which orientation (8ths, 10ths, 12ths, or 16ths on upper edge) the Stanley name and rule number were stamped. Examples with all four possible combinations are common.

30½ B Patternmakers' Shrinkage Rule, 24 1/6 Inch
Offered: 1900 thru 1932

Identical to No. 30½A, except 1/12 inch per foot shrinkage allowance.

30½ C Patternmakers' Shrinkage Rule, 24 1/5 Inch
Offered: 1900 thru 1943

Identical to No. 30½A, except 1/10 inch per foot shrinkage allowance.

30½D Patternmakers' Shrinkage Rule, 24 3/16 Inch
Offered: 1900 thru 1922

Identical to No. 30½A, except 3/32 inch per foot shrinkage allowance.

30½ E Patternmakers' Shrinkage Rule, 24 1/4 Inch
Offered: 1900 thru 1943

Identical to No. 30½A, except 1/8 inch per foot shrinkage allowance.

Figure 79: No. 30½B, 1902–1918 *(Author's Collection)*

Figure 80: No. 30½C, 1900 *(Author's Collection)*

Figure 81: No. 30½E, 1922–1932 *(Author's Collection)*

30½ F Patternmakers' Shrinkage Rule, 24⅜ Inch
Offered: 1900 thru 1943

Identical to No. 30½A, except 3/16 inch per foot shrinkage allowance.

Figure 82: No. 30½F, 1913–1921 *(Author's Collection)*

30½ G Patternmakers' Shrinkage Rule, 24½ Inch
Offered: 1900 thru 1943

Identical to No. 30½A, except ¼ inch per foot shrinkage allowance.

30½ H Patternamkers' Shrinkage Rule, 24 5/16 Inch
Offered: 1912 thru 1932

Identical to No. 30½A, except 5/32 inch per foot shrinkage allowance.

30½ J Patternmakers' Shrinkage Rule, 24 2/5 Inch
Offered: 1912 thru 1922

Identical to No. 30½A, except 1/5 inch per foot shrinkage allowance.

30½ K Patternmakers' Shrinkage Rule, 24 7/16 Inch
Offered: 1912 thru 1932

Identical to No. 30½A, except 7/32 inch per foot shrinkage allowance.

30½ L Patternmakers' Shrinkage Rule, 24 5/8 Inch
Offered: 1912 thru 1943

Identical to No. 30½A, except 5/16 inch per foot shrinkage allowance.

30½ M Patternmakers' Shrinkage Rule, 24¾ Inch
Offered: 1912 thru 1943

Identical to No. 30½A, except 3/8 inch per foot shrinkage allowance.

The Bench Rule

For working at the bench, when compactness and transportability are not important, a folding rule is unnecessary, and can be replaced by a cheaper and stronger non-folding one, the so-called "bench" rule. Such rules are wider, and often thicker than the ordinary 4 fold rule, and are usually 1 or 2 feet in length.

The first two Stanley Bench Rules were the Nos. 34 and 35, both offered from 1855 (Ref. 1) on. These rules were almost identical, both being 2 feet long, 1 5/16 inch wide, and made of maple with brass tips. The only difference between them was that the No. 35 was marked on one side with board measure tables, and the No. 34 was not.

A third bench rule, the No. 31, was introduced in 1874 (Ref. 10) but was discontinued only three years later, between 1877 and 1879 (Ref. 12). This rule was also 2 feet long, but was made of satinwood, instead of maple, and was described as having "brass capped ends," instead of brass tips.

The No. 35 was discontinued after 1892 (Ref. 15). The No. 34 has remained in the line (with minor changes) until the present day (Ref. 68), one of the four surviving traditional rules still made by Stanley. At 130 years, it has had the longest period of continuous production of any Stanley rule.

Two other rules were also sold by Stanley as bench rules, although, as mentioned earlier, they had originally been introduced as "school" rules. These two 1 foot rules, the Nos. 34¼ and 34½, were only named bench rules when they were redesigned in 1934, to have vertical figures, and renumbered as the Nos. 34¼V and 34½V. The No. 34½V was manufactured until 1943 (Ref. 53); the No. 34¼V, up until the present.

31 Bench Rule, 2 Foot
Offered: 1874 thru 1877
Construction: Capped Ends
Material: Satinwood, Brass Caps
Width: Not Known
Graduations: Not Known

It is not clear exactly what the distinction is between the "brass capped ends" of the No. 31, and the conventional brass tips found on most rules; perhaps the caps were cast, and thus could protect the ends of the rule on the edges, as well as on the sides.

No information is available as to the width, thickness, or graduations of the No. 31, but it must have been a premium product; it sold for four times the cost of the equivalant maple bench rule with plain brass tips, the No. 34 (Refs. 10 and 11).

The Folding Shrinkage Rule

The joint in the 2 fold shrinkage rule was unusual in two respects: it was a knuckle joint, instead of the edgewise (rule) joint that was almost universally used for 2 fold rules, and the joint was "triple plated" for strength. This latter description meant that the joint had four sets of plates, two on the edges and two within the wood (sort of a combination of the middle plate and edge plate joints used on the 4 fold rules). Multiple plate knuckle joints were not unique to the 2 fold shrinkage rules (premium rules, such as the No. 72½, were sometimes double or triple plated, as were the longer rules where the joint had to be extra strong, such as the No. 66¼), but they were fairly uncommon.

It is not clear what the function of jointing a shrinkage rule could be. The patternmaker usually worked in a shop at the bench, and would not have much occasion to wish to carry the rule in his pocket. Additionally, his tools were normally kept in a chest, and it would be an unusually small chest that could not hold a collection of nonfolding 2 foot shrinkage rules. It is not surprising that the demand for folding shrinkage rules should be lower than that for the nonfolding variety, and that Stanley should have discontinued their production before 1919 (Ref. 33), while continuing the other type until the 1940's.

31 Patternmakers' Shrinkage Rule, 24¼ Inch, 2 Fold
Offered: 1884 thru 1910
Construction: Knuckle Joint, Triple Plated Edge Plates
Material: Boxwood, Brass Hinge & Tips
Width: 1 5/16 Inch
Graduations: 8ths and 16ths of Inches @ 1/8 Inch per Foot Shrinkage

The No. 31 was another of the four rules which were retained when Stanley discontinued the rest of the Stearns line of wood rules in 1884 (Ref. 13) (the others were the Nos. 13, 30, and 80). Stanley had not offered shrinkage rules prior to that time, and the catalogue description of this "new" rule corresponded exactly with that of the Stearns No. 82, which was listed for the last time in 1879 (Ref. 12). It is highly probable that this was not a coincidence, but was simply the transfer to the Stanley product line of a rule too popular to be eliminated along with the Stearns name.

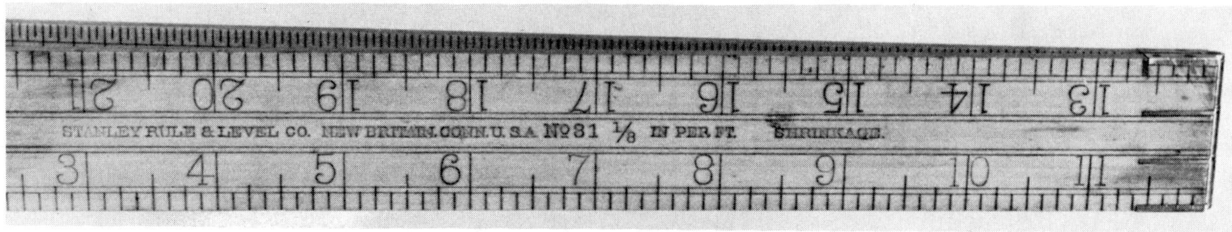

Figure 83: No. 31, 1884–1910 *(Author's Collection)*

Like its counterpart, the No. 30, the No. 31 was initially offered only in a single shrinkage allowance, 1/8 inch per foot. In 1898, again like the No. 30, the number of shrinkage allowances available was increased to six, but with the five new allowances 1/16, 1/10, 3/32, 3/16, and 1/4 inch per foot) only available on special order. In 1900, with the introduction of the new Nos. 30½ and 31½ shrinkage rules, the offer of special-order allowances for the No. 31 was withdrawn, and within eleven more years, before 1912, the rule itself was discontinued.

31-1/2A Patternmakers' Shrinkage Rule, 24 1/8 Inch, 2 Fold
Offered: 1902 thru 1917
Construction: Knuckle Joint, Triple Plated Edge Plates
Material: Boxwood, Brass Hinge & Tips
Width: 1½ Inch
Graduations: 8ths, 10ths, 12ths, and 16ths of Inches @ 1/16 Inch per Foot Shrinkage

The No. 31½ folding shrinkage rule occupies the same position vis-a-vis its predecessor, the No. 31, that the No. 30½ did to the No. 30: it was essentially a redesign, being wider, and having more graduations.

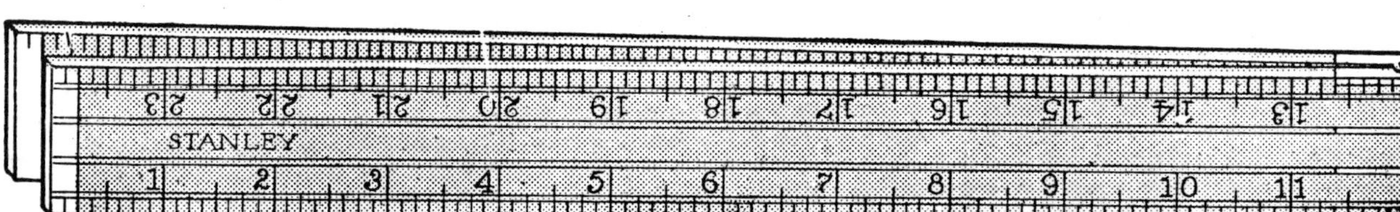

Figure 84a: No. 31½A, 1915 *(Ref. 31)*

The No. 31½ did not have a long lifetime; it was introduced in 1902 (Ref. 20), but was discontinued between 1917 and 1919 (Refs. 32 and 33). During that short period, however, it was produced in all of the shrinkage allowances available for the No. 30½. The No. 31½ was initially introduced in seven different shrinkage allowances, a number that was ten years later increased to 12 (Ref. 27). Again, as with the No. 30½, from 1909 on the shrinkage allowance was indicated by a letter suffix appended to the rule number in the catalogue. The shrinkage allowances, together with the associated suffix, and the years offered are shown in the following table:

Shrinkage Allowance	Letter Suffix	Years Offered
1/16 Inch/Foot	A	1902–1917
1/12 Inch/Foot	B	1902–1917
3/32 Inch/Foot	D	1902–1917
1/10 Inch/Foot	C	1902–1917
1/8 Inch/Foot	E	1902–1917
5/32 Inch/Foot	H	1912–1917
3/16 Inch/Foot	F	1902–1917
1/5 Inch/Foot	J	1912–1917
7/32 Inch/Foot	K	1912–1917
1/4 Inch/Foot	G	1902–1917
5/16 Inch/Foot	L	1912–1917
3/8 Inch/Foot	M	1912–1917

Figure 84b: Table of Letter Suffixes of No. 31½ Rules

The letter suffix was not initially marked on the rule, only the rule number (31½) and the shrinkage factor. Later, some time after 1912, this was changed, and both the suffix and the shrinkage allowance were stamped.

Throughout its lifetime the No. 31½ was available graduated in 8ths and 16ths only, at no extra cost.

31½ B Patternmakers' Shrinkage Rule, 24 1/6 Inch, 2 Fold
Offered: 1902 thru 1917

Identical to No. 31½A, Except 1/12 Inch per Foot Shrinkage

31½ C Patternmakers' Shrinkage Rule, 24 1/5 Inch, 2 Fold
Offered: 1902 thru 1917

Identical to No. 31½A, Except 1/10 Inch per Foot Shrinkage

31½ D Patternmakers' Shrinkage Rule, 24 3/16 Inch, 2 Fold
Offered: 1902 thru 1917

Identical to No. 31½A, Except 3/32 Inch per Foot Shrinkage

31½ E Patternmakers' Shrinkage Rule, 24 1/4 Inch, 2 Fold
Offered: 1902 thru 1917

Identical to No. 31½A, Except 1/8 Inch per Foot Shrinkage

31½ F Patternmakers' Shrinkage Rule, 24 3/8 Inch, 2 Fold
Offered: 1902 thru 1917

Identical to No. 31½A, Except 3/16 Inch per Foot Shrinkage

31½ G Patternmakers' Shrinkage Rule, 24½ Inch, 2 Fold
Offered: 1902 thru 1917

Identical to No. 31½A, Except ¼ Inch per Foot Shrinkage

31½H Patternmakers' Shrinkage Rule, 24$^{5}/_{16}$ Inch, 2 Fold
Offered: 1912 thru 1917

Identical to No. 31½A, Except $^{5}/_{32}$ Inch per Foot Shrinkage

(Author's Collection)

Figure 85: No. 31½H, 1905–1917

31½J Patternmakers' Shrinkage Rule, 24$^{2}/_{5}$ Inch, 2 Fold
Offered: 1912 thru 1917

Identical to No. 31½A, Except $^{1}/_{5}$ Inch per Foot Shrinkage

31½ K Patternmakers' Shrinkage Rule, 24$^{7}/_{16}$ Inch, 2 Fold
Offered: 1912 thru 1917

Identical to No. 31½A, Except $^{7}/_{32}$ Inch per Foot Shrinkage

31½ L Patternmakers' Shrinkage Rule, 24⅝ Inch, 2 Fold
Offered: 1912 thru 1917

Identical to No. 31½A, Except $^{5}/_{16}$ Inch per Foot Shrinkage

31½ M Patternmakers' Shrinkage Rule, 24¾ Inch, 2 Fold
Offered: 1912 thru 1917

Identical to No. 31½A, Except ⅜ Inch per Foot Shrinkage

32 Carpenters' Caliper Rule, 1 Foot, 4 Fold

Offered: 1859 thru 1941

Construction: Arch Joint, Edge Plates
Left Hand Caliper (1932 and before)
Right Hand Caliper (1934 and after)

Material: Boxwood, Brass Hinges, Caliper, & Tips

Width: 1 Inch

Graduations: Body: 8ths, 10ths, and 16ths of Inches (1872 and before)
8ths, 10ths, 12ths, and 16ths of Inches (1874 and after)
Caliper: 16ths of Inches (1922 and before)
16ths and 32nds of Inches (1925 and after)

Between 1888 (Ref. 14) and 1892 (Ref. 15) the caliper jaw on this rule was changed slightly, to have a rounded back (Fig. 86b) instead of the hollow back (Fig. 86a) which had been the style until that time. No obvious reason can be found for this change; it must have been made for merely cosmetic purposes, or possibly to slightly reduce manufacturing cost (the hollow would have had to be milled, but the rounded back could be satisfactorily shaped by hand on a grinding disk).

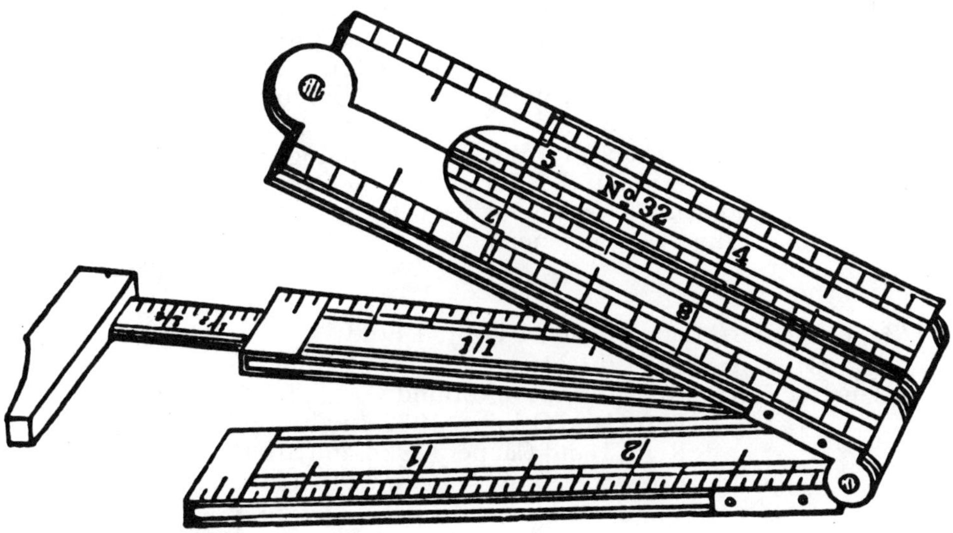

Figure 86a: No. 32, 1867 *(Ref. 7)*

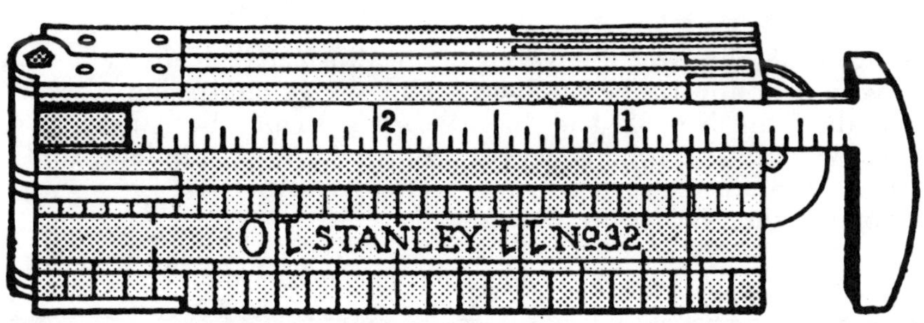

Figure 86b: No. 32, 1919 *(Ref. 33)*

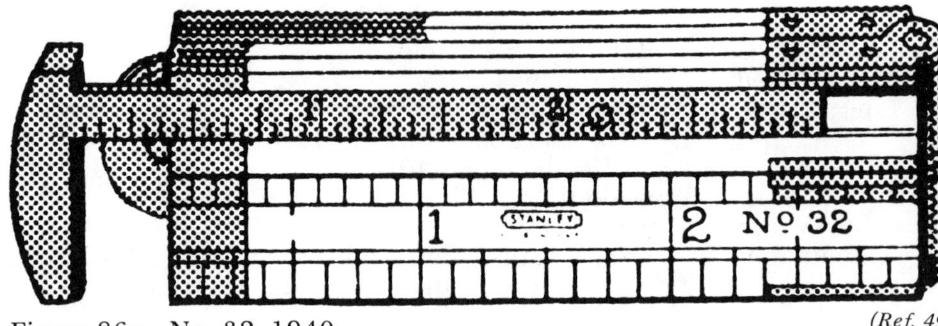

Figure 86c: No. 32, 1940 *(Ref. 49)*

This design change was not confined to the No. 32; the No. 38, the only other Stanley caliper rule for which there exists a pre-1892 catalogue illustration, was similarly modified at the same time. Rules wider than 1 inch (such as the Nos. 13 and 36½) were not so modified; it would seem that Stanley had decided that all of the narrow caliper rules would have a rounded jaw, and all of the wider a hollow one. This is only a hypothesis, but the examination of post-1892 catalogue illustrations and the study of numerous examples would seem to support it.

In 1934 the No. 32 was changed from having a left hand caliper (see SLIDES AND CALIPERS Chapter 2) to having a right hand caliper (see Fig. 86c). This was a general change, made at this time to all Stanley caliper rules still in production, and was probably intended to make these rules more acceptable in the English and European markets, where right hand caliper rules were the norm.

In 1942 the No. 32, along with many other rules, was discontinued to conserve materials and to make room for more important war production. At that time the price list (Ref. 52) described this cancellation as only "temporary," but this decision was subsequently modified, and when the first postwar catalogue was published in 1947 (Ref. 54) the No. 32 had been dropped permanently.

32-1/2 Carpenters' Caliper Rule, 1 Foot, 4 Fold

Offered:	1874 thru 1941	
Construction:	Arch Joint, Full Bound	
	Left Hand Caliper (1932 and before)	
	Right Hand Caliper (1934 and after)	
Material:	Boxwood, Brass Hinges, Caliper, & Trim	
Width:	1 Inch	
Graduations:	Body:	8ths, 10ths, 12ths, and 16ths of Inches
	Caliper:	16ths of Inches (1922 and before)
		16ths and 32nds of Inches (1925 and after)

From 1874 to 1888 the No. 32½ probably was made with a hollow-backed caliper jaw, similar to that of the No. 32, and, like that rule, was converted to a round-backed jaw shape between 1888 and 1892. This cannot be known for certain; the No. 31½ was not illustrated in any catalogue prior to 1909, and no examples have been found with the earlier shape, but these two rules differed only in their binding, and were probably identical otherwise.

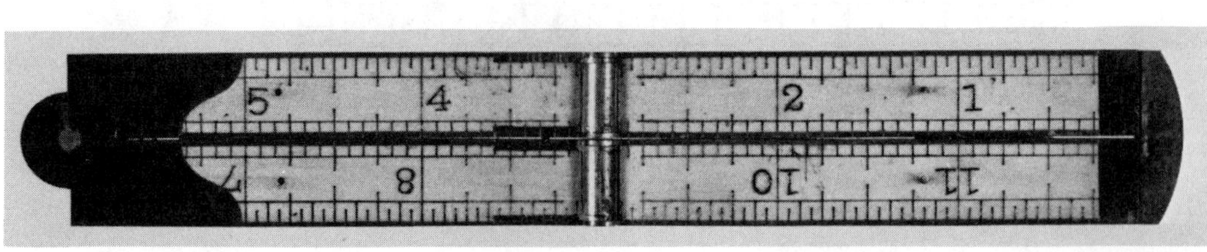

Figure 87a: No. 32½, 1892—1921 *(Author's Collection)*

In 1934 (Ref. 44) the No. 32½, as in the case of the other Stanley caliper rules, was changed from having a left hand caliper (see SLIDES AND CALIPERS, Chapter 2) to having a right hand caliper (see Fig. 87b). This was a general change, made at this time to all Stanley caliper rules still in production, and was probably intended to make these rules more acceptable in the English and European markets, where right hand caliper rules were the norm.

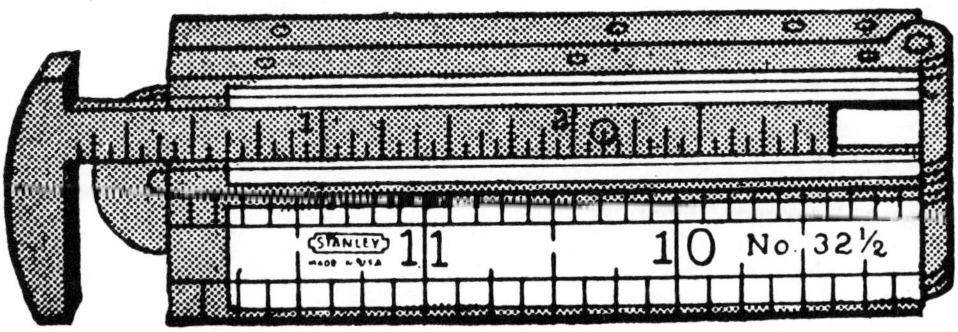

Figure 87b: No. 32½, 1934 *(Ref. 44)*

The discontinuance of the No. 32½ was similar to that of the No. 32. "Temporarily" discontinued in 1942 (Ref. 52), it was never reintroduced after the end of the war, one of the thirty-one rules which fell victim to war production and postwar retrenchment.

The Yard Stick

The yard stick is probably the most common measuring instrument encountered in everyday life, and has been for many years. Its 3 foot length makes it adequate for most measuring tasks around the home, and its dual graduations (fractions of an inch and fractions of a yard) adapt it to almost any measuring situation.

It is the scale of fractions of a yard that distinguishes the yard stick from the common 3 foot rule or stick. This scale (usually to 8ths of a yard) allows the stick to be used for measuring fabric for tailoring purposes (sewing patterns traditionally specify the amount of fabric required in yards, and that is the way it is sold from the bolt). Many stores dealing in "yard goods" would not only have several yard sticks loose on hand, but would often have one or more fastened to the counter, to prevent loss, and facilitate use.

Yard sticks have formed part of the Stanley product line from the beginning, and have been one of the most stable and least changing class of rules of any. Two sticks, the Nos. 41 (maple) and 66 (boxwood, 4 fold), both 1 inch wide, were offered in 1855 (Ref. 1). These were supplemented by two additional narrower (¾ inch) sticks in 1867 (Ref. 7), the Nos. 33 (maple, untipped) and 50 (hickory). For the next fifty years, until the time of World War I, these four rules, unchanged, constituted the Stanley line of yard sticks.

The No. 66 was dropped from the product line after 1917 (Ref. 32). The Nos. 33 and 50 were continued in production for another twenty-five years until 1941–1942 (Refs. 51 and 52), at which time they were also discontinued. The No. 41, renamed the No. 41YS in 1961 (Ref. 62), was only dropped after 1972 (Ref. 68), after being produced essentially unchanged for more than 115 years.

Initially both scales on Stanley yardsticks were graduated from right to left, consistent with the graduations on 2 fold and 4 fold rules. About 1925, however, this was reversed, and all yard sticks made after this date were graduated from left to right. It is not known why this change was made; for some reason there seems to have been a general trend toward left to right graduations at about this time, and the yard sticks had been included.

During this period, only two attempts, neither notably successful, were made to expand the line of yard sticks.

In 1925 Stanley began marketing under its own name the Nos. 450 and 550 counter measures hitherto manufactured by the recently acquired Eagle Square Manufacturing Co. These were thin steel yard measures, intended for use fastened to a counter in a retail store, even to the extent of having predrilled nail holes. Both were dropped from the product line after 1932 (Ref. 43), probably as a result of declining sales due to the depression of 1929–1939.

In 1940 (Ref. 49), Stanley added another new yard stick to the line, the No. 214T. This was a cheaper stick than the others, made of softwood, with printed figures and graduations and steel tips. A second, and probably similar, yard stick in this class, the No. 214, was added in 1941 (Ref. 51), but almost immediately discontinued. The No. 214T itself was dropped some time between 1943 (Ref. 53) and 1947 (Ref. 54).

33 Yard Stick, 3 Foot

Offered:	1860 thru 1941
Construction:	Polished
Material:	Maple
Width:	¾ Inch
Graduations:	8ths of Inches
	Fractions of a Yard

In 1942 (Ref. 52) the No. 33 was "temporarily" discontinued to make room for more important war production. This temporary hiatus proved to be permanent, however, and this yard stick did not reappear in the catalogue after the end of hostilities (Ref. 54).

Figure 88: No. 33, 1922–1932 *(Author's Collection)*

34 Bench Rule, 2 Foot

Offered:	1855 thru 1932
Material:	Maple, Brass Tips
Width:	$1\frac{5}{16}$ Inch (1909 and before)
	$1\frac{1}{4}$ Inch (1912 and after)
Graduations:	8ths and 16ths of Inches

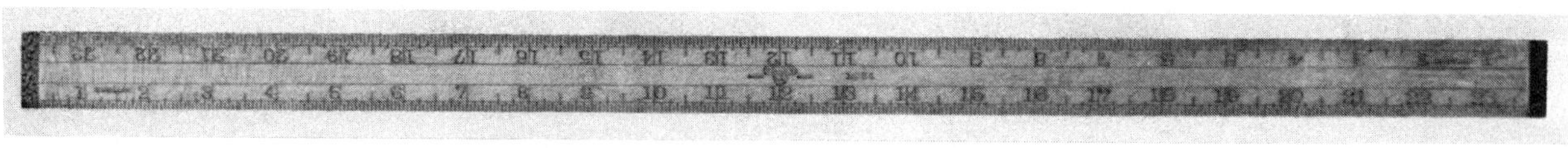

Figure 89: No. 34, 1922–1932 *(Charles and Walter Jacob Collection)*

34V Bench Rule, 2 Foot

Offered:	1934 thru 1960
Construction:	Vertical Figures
Material:	Maple, Brass Tips
Width:	1¼ Inch
Graduations:	8ths and 16ths of Inches

The No. 34V is identical to the No. 34 except for having vertical, instead of horizontal, figures.

34VR Bench Rule, 2 Foot

Offered:	1961 thru 1975

Identical to the No. 34V; a model number change only.

34-1/4 School Rule, 1 Foot

Offered:	1919 thru 1932
Material:	Maple, Brass Tips
Width:	1¼ Inch (1922 and before)
	1⅛ Inch (1925 and after)
Graduations:	8ths and 16ths of Inches

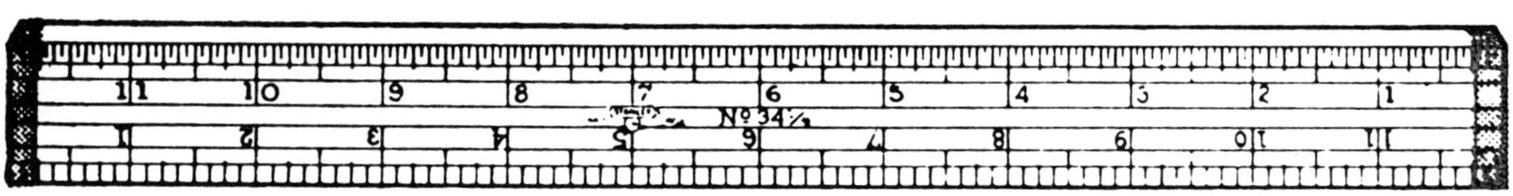

Figure 90: No. 34¼, 1929 *(Ref. 41)*

34-1/4V Bench Rule, 1 Foot

Offered:	1934 thru 1942, 1947 thru 1960
Construction:	Vertical Figures
Material:	Maple, Brass Tips
Width:	1⅛ Inch
Graduations:	8ths and 16ths of Inches

The No. 34¼V is identical to the No. 34¼ except for having vertical, instead of horizontal, figures.

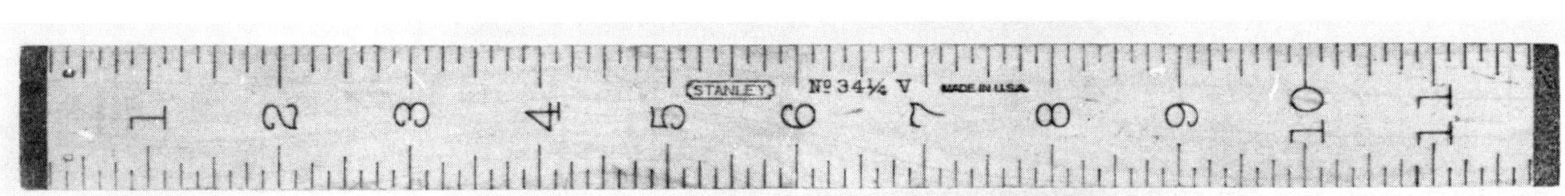

Figure 91: No. 34¼V, 1934–1960 *(Author's Collection)*

In 1943, the No. 34¼V, along with a few others, was added to the list of rules canceled to further conserve materials and make room for increased war production. The price list (Ref. 53) did not identify this cancellation as either "permanent" or "temporary," but it must have been the latter; when the first postwar catalogue was published in 1947 (Ref. 54) the No. 34¼V was back in the product line.

34-1/4VR Bench Rule, 1 Foot
Offered: 1961 thru 1975

Identical to No. 34¼V; a model number change only

34-1/2 School Rule, 1 Foot
Offered: 1919 thru 1932
Material: Boxwood, Brass Tips
Width: 1¼ Inch (1922 and before)
1⅛ Inch (1925 and after)
Graduations: 8ths and 16ths of Inches

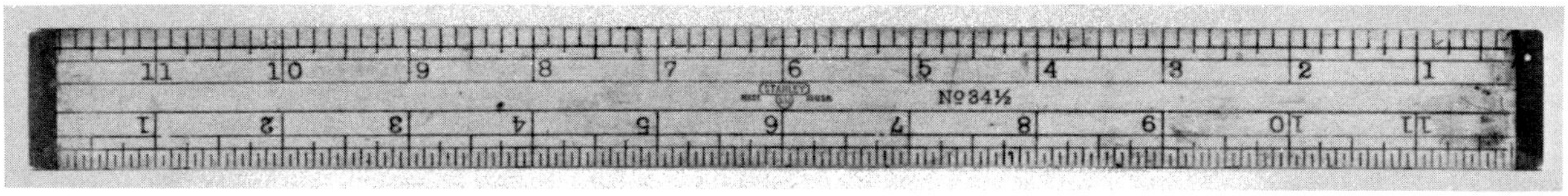

Figure 92: No. 34½, 1922–1932 *(Author's Collection)*

34-1/2V Bench Rule, 1 Foot
Offered: 1934 thru 1943
Construction: Vertical Figures
Material: Boxwood, Brass Tips
Width: 1⅛ Inch
Graduations: 8ths and 16ths of Inches

The No. 34½V is identical to the No. 34½ except for having vertical, instead of horizontal, figures.

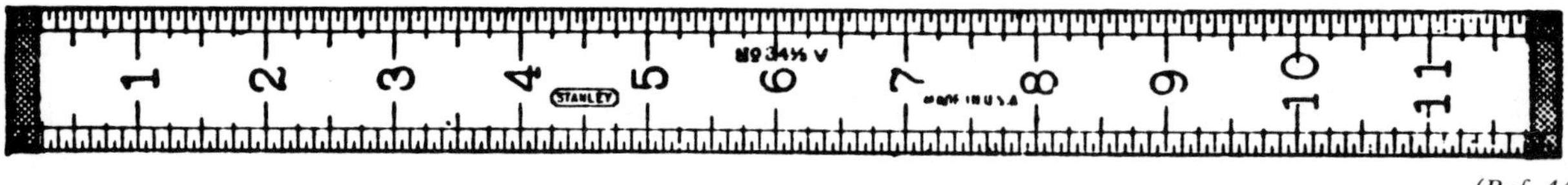

Figure 93: No. 34½V, 1934 *(Ref. 44)*

Board Measure and Board Sticks

A common method of measuring gross lumber quanitites is to reduce the volumes of the various pieces in a lot to the equivalent number of square feet of lumber one inch thick, and to refer to their sum as the number of "Board Feet" in the lot. Wood is frequently priced and sold by the board foot, and estimates of the amount of wood required for a job are often prepared in these units.

The process of measuring lumber and computing the number of board feet is not a difficult one, but it is time consuming, and subject to computational error. It is not surprising, therefore, that Stanley should manufacture special rules with tables or scales that would aid the woodworker in this task.

Two types of such rules were offered: ordinary rules imprinted with board measure tables, and "board sticks," specially graduated with board scales.

Board Measure Tables were simply that: tables of board foot values for boards 1 inch thick, having as their two axis board lengths in feet and board widths in inches. On Stanley rules these were actually two tables arranged side by side: a small table for boards 1 to 12 feet long and 6 to 12 inches wide (see Figure 94), and, to its right, a larger table for boards 1 to 19 feet long and 13 to 19 inches wide (see Figure 95).

	1	2	3	4	5	6	7	8	9	10	11	12
6	.6	1.	1.6	2.	2.6	3.	3.6	4.	4.6	5.	5.6	6.
7	.7	1.2	1.9	2.4	2.11	3.6	4.1	4.8	5.3	5.10	6.5	7.
8	.8	1.4	2.	2.8	3.4	4.	4.8	5.4	6.	6.8	7.4	8.
9	.9	1.6	2.3	3.	3.9	4.6	5.3	6.	6.9	7.6	8.3	9.
10	.10	1.8	2.6	3.4	4.2	5.	5.10	6.8	7.6	8.4	9.2	10.
11	.11	1.10	2.9	3.8	4.7	5.6	6.5	7.4	8.3	9.2	10.1	11.
12	1.	2.	3.	4.	5.	6.	7.	8.	9.	10.	11.	12.

Figure 94: Small Board Measure Table

When placed on folding rules these tables were marked half on each leg (on the inside surface in the case of 4 fold rules), and thus could only be used when the main joint was closed. Furthermore, although they were stamped in extremely small figures, they were still physically quite large, and were only placed on rules at least $1^{5}/_{16}$ inches wide (the No. 35 bench rule, the Nos. 7 and 22 2 fold rules, and the Nos. 79, 80, 81, and 82 4 fold rules).

The tables were used by rounding the measured board length and width to the nearest units, and then looking up the value in the table at that column and row. The result was a number of board feet and twelfths of board feet ("board inches"), separated by a decimal point. Extrapolation could be used to improve accuracy slightly, but most users did not bother.

These ordinary rules with board measure tables were primarily intended for occasional use. In commercial applications considerations of speed and accuracy made it more common to use the so-called board stick, with its set of board measure scales. These were "computing" scales, each scale reading directly in board feet for a 1 inch thick board of a different length. Knowing the length of a board, the stick would be laid across it, and the number of board feet read off the appropriate scale without the need for a table lookup. These board scales took up all the space on the stick, leaving no room for ordinary 8ths and 16ths scales (although the markings on the scale for 12 foot boards, being 1 inch apart, could be used as a substitute), and thus these were special purpose tools intended only for lumber measurement.

At different times at least eleven different board sticks were offered, in a wide variety of designs. Either 2 feet or 3 feet in length, they were always provided with a hook or lipped head at one end, to allow one-handed use, and graduated left to right beginning at that end. Some, the Nos. 43, 43½, and 49, were flat in cross section; others, the Nos. 44, 45, 46½, and 47½ (1870–1915),

	1	2	3	4	5	6	7	8	9	10	11	12	13	14	15	16	17	18	19
13	1.1	2.2	3.3	4.4	5.5	6.6	7.7	8.8	9.9	10.10	11.11	13.	14.1	15.2	16.3	17.4	18.5	19.6	20.7
14	1.2	2.4	3.6	4.8	5.10	7.	8.2	9.4	10.6	11.8	12.10	14.	15.2	16.4	17.6	18.8	19.10	21.	22.2
15	1.3	2.6	3.9	5.	6.3	7.6	8.9	10.	11.3	12.6	13.9	15.	16.3	17.6	18.9	20.	21.3	22.6	23.9
16	1.4	2.8	4.	5.4	6.8	8.	9.4	10.8	12.	13.4	14.8	16.	17.4	18.8	20.	21.4	22.8	24.	25.4
17	1.5	2.10	4.3	5.8	7.1	8.6	9.1	11.4	12.9	14.2	15.7	17.	18.5	19.10	21.3	22.8	24.1	25.6	26.11
18	1.6	3.	4.6	6.	7.6	9.	10.6	12.	13.6	15.	16.6	18.	19.6	21.	22.6	24.	25.6	27.	28.6
19	1.7	3.2	4.9	6.4	7.11	9.6	11.1	12.8	14.3	15.10	17.5	19.	20.7	22.2	23.9	25.4	26.11	28.6	30.1

Figure 95: Large Board Measure Table

were square, while still others, the Nos. 46, 47, 47½ (1865), and 48, were octagonal. The number of scales on a stick varied as a function of its cross-sectional shape and circumference; some having as few as six, and others as many as sixteen. Two, the Nos. 47½ (1865) and 48, were even shaped like canes; they were octagonal, tapering from ¾ inch at the head to less than ½ inch at the tip, which was equipped with a brass ferrule (see Figure 117).

A total of twelve board sticks and rules with board tables were listed in the 1855 price list of A. Stanley & Co. (Ref. 1). Seven of these were rules with board tables, the Nos. 7, 22, 35, 79, 80, 81, and 82, and five with board scales (the Nos. 44, 45, 46, 47, and 48).

Of the seven rules with board tables, two, the Nos. 7 and 80, had only a short history, and were eliminated from the product line sometime between 1859 (Ref. 2) and 1862 (Ref. 5). The other five, the Nos. 22, 35, 79, 81, and 82, were better received, and continued to be sold until 1915–1917 (Refs. 31 and 32).

The line of board sticks was more dynamic, and subject to more frequent additions and deletions. Two of the five, the Nos. 44 and 45 (square in cross-section), had been dropped by the time the 1859 price list (Ref. 2) was issued and three new ones (flat in cross-section) added: the No. 43 (a very short-lived stick; it was only listed in 1859 and 1860 [Ref. 3]), and the Nos. 43½ and 49. A second board cane, the No. 47½, was offered in 1865 (Ref. 6), only to be discontinued again before 1867 (Ref. 7). In 1872 (Ref. 9), with the addition of the Nos. 46½ and 47½, square board sticks were reintroduced to the line; apparently the cancellation of the Nos. 44 and 45 in 1855–1859 had been premature, and there had been more demand for a square stick than had been realized.

None of the eleven board sticks and rules with board measure tables which were still offered in 1915 were to survive America's entry into the First World War. By that time, the Lufkin Rule Co., of Saginaw, Michigan, had taken Stanley's place as the preeminent manufacturer of log rules and board sticks in the United States. Additionally, the cost and scarcity of Turkey boxwood and the consequent switch to Maracaibo boxwood had significantly reduced the profit margins on most folding rules, including the Nos. 22, 79, 81, and 82. Thus when the need arose to divert manufacturing resources to war production, and many rules were discontinued, these rules, sticks, and canes were among them. Ten of the eleven were listed for the last time in 1915 (Ref. 31). The eleventh, the No. 82, was for some reason retained for two more years (Ref. 32) before it was also canceled (possibly to use up existing stocks).

35 Bench Rule, 2 Foot

Offered:	1855 thru 1892
Material:	Satinwood (1855 thru 1860) Maple (1862 and after) Brass Tips (1874 and after)
Width:	1 5/16 Inch
Graduations:	8ths and 16ths of Inches
Other Markings:	Board Measure Tables

The No. 35 was a variant of the No. 34 Bench Rule, with some of the scales omitted to make room for board measure tables. The two rules were the same length and width, and were graduated identically on one side in 8ths and 16ths of inches. On the other side, however, where the No. 34 had the same graduations repeated, the No. 35 was only graduated on its left hand half, with the right hand half being marked instead with board measure tables (see Fig. 96).

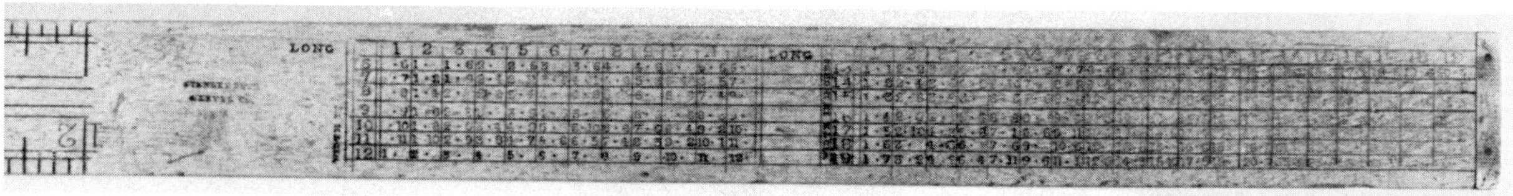

Figure 96: No. 35, 1874–1892 *(Author's Collection)*

The tables marked on this rule were identical to those on 2 fold and 4 fold rules: a small table for 6 to 12 inch boards, and, to its right, a larger one for 13 to 19 inch boards.

In 1855 (Ref. 1), and later in 1860 (Ref. 3) this bench rule was described as being made of "Satin Wood," but in 1859 (Ref. 2), and from 1862 (Ref. 5) on, no such statement was made. West Indian Satinwood *(Zanthoxylum flavum)* was the wood referred to. Probably the supply was interrupted by the Civil War, necessitating the use of maple instead, and after the end of hostilities it was found simpler to continue with that commoner wood.

35 Carpenters' Caliper Rule, 1 Foot, 2 Fold

Offered:	1910 thru 1917	
Construction:	Square Joint, Left Hand Caliper	
Material:	Boxwood, Brass Hinge, Caliper, & Tips	
Width:	1 Inch	
Graduations:	Body:	8ths, 10ths, 12ths, and 16ths of Inches
	Caliper:	16ths of Inches

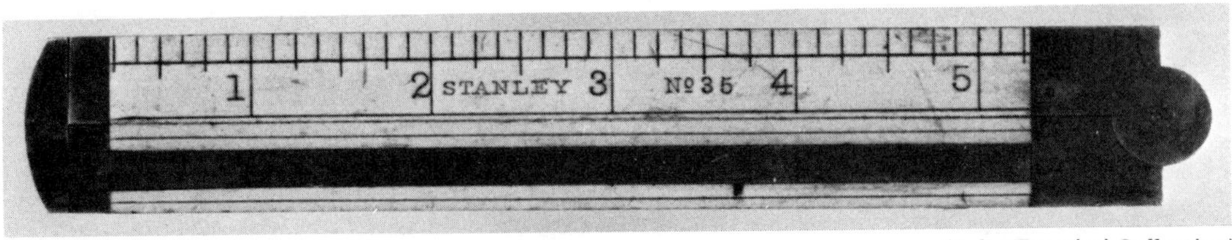

Figure 97: No. 35, 1910–1917 *(John Treggiari Collection)*

36 Carpenters' Caliper Rule, 6 Inch, 2 Fold

Offered:	1855 thru 1941, 1947 thru 1949	
Construction:	Square Joint Left Hand Caliper (1932 and before) Right Hand Caliper (1934 and after)	
Material:	Boxwood, Brass Hinge, Caliper & Trim	
Width:	7/8 Inch (1936 and before) 1 Inch (1938 and after)	
Graduations:	Body:	8ths and 16ths of Inches (1872 and before) 8ths, 10ths, and 16ths of Inches (1874 thru 1877) 8ths, 10ths, 12ths, and 16ths of Inches (1879 and after)
	Caliper:	16ths of Inches (1922 and before) 16ths and 32nds of Inches (1925 and after)

In the form in which the No. 36 was originally offered, the surface of the leg opposite that containing the caliper slide was not graduated, but was simply marked with the Stanley trademark and the rule number (Fig. 98a). Sometime between 1907 (Ref. 23) and 1909 (Ref. 25) this was changed, however, and that surface was from then on graduated in 8ths of inches (Fig. 98b). At the same time, the other markings on that side were reduced to the rule number and the name STANLEY.

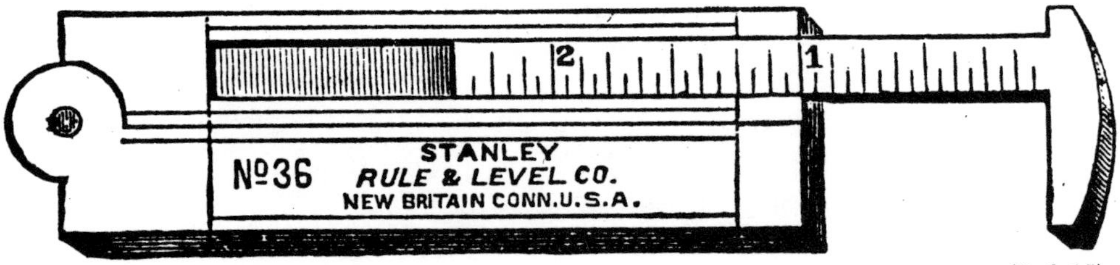

Figure 98a: No. 36, 1892 *(Ref. 15)*

Figure 98b: No. 36, 1922–1932 *(Eugene Frankio Collection)*

In 1934 the No. 36 was changed from having a left hand caliper (see SLIDES AND CALIPERS, Chapter 2) to having a right hand caliper (see Figure 98c). This was a general change, made at this time to all Stanley caliper rules still in production, and was probably intended to make these rules more acceptable in the English and European markets, where right hand caliper rules were the norm.

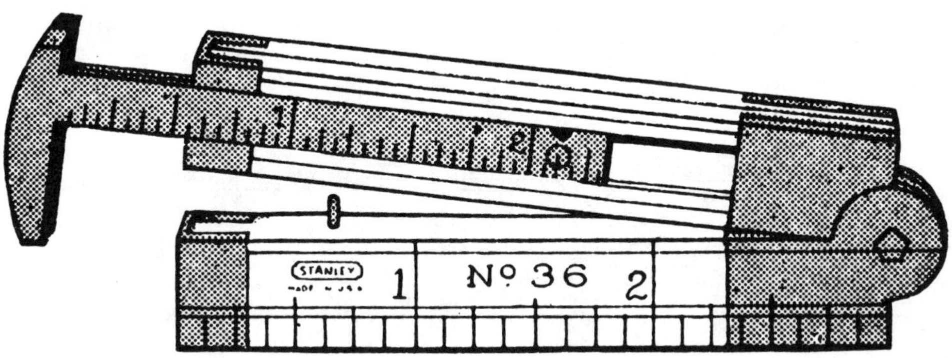

Figure 98c: No. 36, 1934 *(Ref. 44)*

In 1942 the No. 36, along with many other rules, was discontinued to conserve materials and make room for more important war production. The price list (Ref. 52) described this cancellation as "temporary," and although some other rules so canceled were never returned to production this was not so with the No. 36. It was offered in the first postwar catalogue in 1947 (Ref. 54), and two years later in 1949 (Ref. 55), before finally being permanently dropped.

36-1/2 Carpenters' Caliper Rule, 1 Foot, 2 Fold

Offered: 1855 thru 1859, 1874 thru 1940

Construction: Square Joint
Left Hand Caliper (1932 and before)
Right Hand Caliper (1934 and after)
Bitted (1926 thru 1932)

Material: Boxwood, Brass Hinge, Caliper, & Trim

Width: 1$^{3}/_{8}$ Inch

Graduations: Body: 8ths, 10ths, 12ths, and 16ths of Inches
Caliper: 16ths of Inches (1922 and before)
16ths and 32nds of Inches (1925 and after)

The No. 36½ 1 Foot, 2 Fold Caliper Rule was first offered in 1855 (Ref. 1), and was still in production (as the Nos. 36½L and 36½R) as late as 1963 (Ref. 63). This would make it a member of the small group of Stanley rules which were in production over 100 years (the others are the No. 34 Bench Rule, the No. 41 Yard Stick, and the Nos. 61 and 68 4 Fold Rules), except for the fact that it was dropped from the catalogue in 1860 (Ref. 3), and not offered again until 1874 (Ref. 10). This twelve year hiatus is curious when one considers how popular this rule must have been to be retained in the product line until the 1960's. While no explanation has been found for it, the most likely reason is that Stanley misjudged the demand for this rule, and then reintroduced it again later when the mistake was recognized.

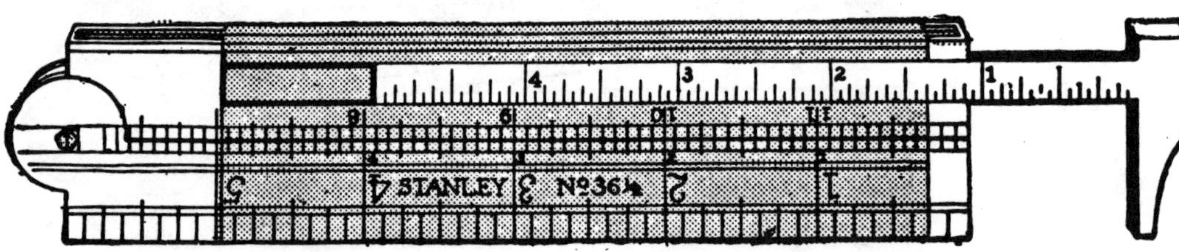

Figure 99a: No. 36½, 1919 *(Ref. 33)*

In 1934 the No. 36½ was changed from having a left hand caliper (see SLIDES AND CALIPERS, Chapter 2) to having a right hand caliper (see Figure 99b). This was a general change, made at this time to all Stanley caliper rules still in production, and was probably intended to make these rules more acceptable in the English and European markets, where right hand caliper rules were the norm.

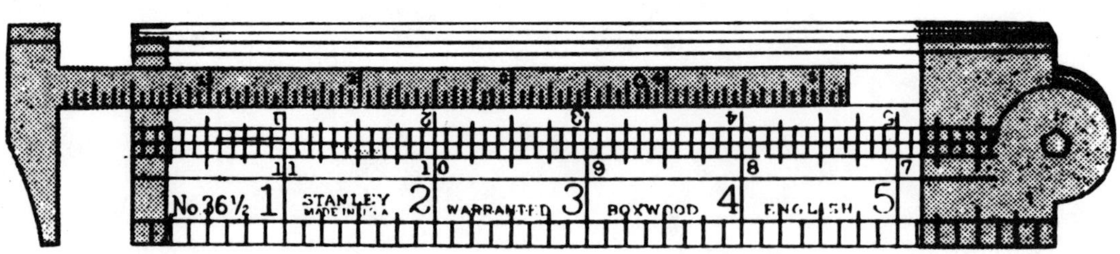

Figure 99b: No. 36½, 1934 *(Ref. 44)*

Between 1940 (Ref. 49) and 1941 (Ref. 50) the left hand caliper version of the No. 36½ was reintroduced, as the No. 36½L. At the same time the right hand caliper version made since 1934 was renumbered as the No. 36½R, and both configurations offered in parallel from then on.

The Nos. 36½L and 36½R were the last boxwood rules to be produced by Stanley; both were still offered as late as 1963 (Ref.

63), three years after the Nos. 61 and 68 4 fold rules were listed for the last time in 1958 (Ref. 60).

36-1/2L Carpenters' Caliper Rules, 1 Foot, 2 Fold
Offered: 1941, 1947 thru 1963

The No. 36½L was a reintroduction of the left hand caliper version of the No. 36½ 1 Foot, 2 Fold rule, the version which had been discontinued when Stanley switched all their caliper rules from left hand to right hand in 1934. Apparently not all users liked the right hand caliper arrangement, and in the case of the extremely popular No. 36½ (it was the last of the folding rules to be discontinued after World War II) there was enough demand to justify offering both configurations. The No. 36½L did not replace the No. 36½R, but was offered in parallel with it, the only known case of Stanley ever offering both the left hand and the right hand versions of a caliper rule at the same time.

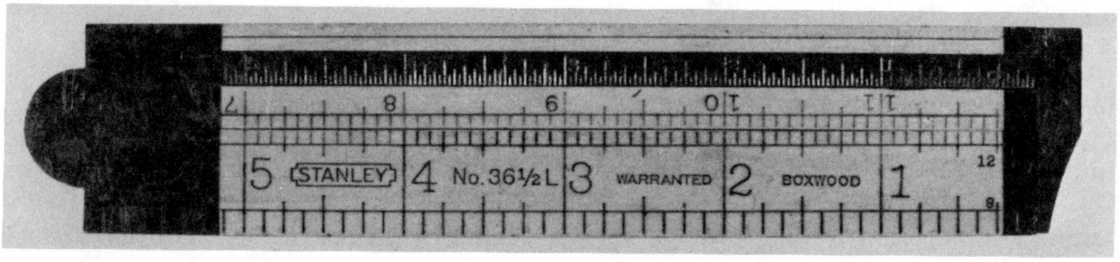

Figure 100: No. 36½L, 1941–1963 *(Author's Collection)*

In 1942, the No. 36½L, along with many other rules, was discontinued to conserve materials and to make room for more important war production. The 1942 Price List (Ref. 52) described this cancellation as only "temporary," and although some other rules so canceled were never returned to production, this was not the case with the No. 36½L, and it was back in the product line when the first postwar catalogue was published in 1947 (Ref. 54).

36-1/2R Carpenters' Caliper Rule, 1 Foot, 2 Fold
Offered: 1941, 1947 thru 1963

The No. 36½R was identical to the No. 36½ offered in 1940; this was a model number change only, for inventory and accounting purposes.

In 1942, the No. 36½R, along with many other rules, was discontinued to conserve materials and to make room for more important war production. The 1942 Price List (Ref. 52) described this cancellation as only "temporary," and although some other rules so canceled were never returned to production, this was not the case with the No. 36½R, and it was back in the product line when the first postwar catalogue was published in 1947 (Ref. 54).

37 Carpenters' Caliper Rule, 6 Inch, 2 Fold

Offered: 1855 thru 1859
Construction: Square Joint, Left Hand Caliper
Material: Ivory, Brass Hinge, Caliper, & Tips
Width: Not Known, but probably 7/8 Inch
Graduations: Not Known

Gauging and Wantage Rods

The gauging rod and the wantage rod are the tools of the gauger; used to measure the capacity and contents of wooden casks. The gauging rod is used to measure the capacity of the cask in gallons; the wantage rod to measure how many gallons the cask lacks of being full.

Casks, of whatever nominal size, from the half-barrel of 16 gallons to the pipe of 120 gallons, are made on the basis of wine measure; that is to say, their proportions (i.e., the ratio of length to head diameter, and the ratio of maximum diameter to head diameter) are always the same. Thus, by measuring a single dimension of a cask, it is possible to calculate its total capacity. The dimension used for this purpose is the distance from the bung hole, in the middle of the side of the barrel, to the opposite chine (the point at which the staves contact the head). This dimension is used for two reasons: 1) i⁺ is capable of easy measurement with a rod inserted through the bung hole, and 2) it represents a good working average of the length, head diameter, and maximum diameter of the cask, thus compensating to a large extent for any manufacturing variation in cask proportions.

The gauging rod is one of the common devices used by the gauger to measure this dimension and convert it into the equivalent capacity. Three or four feet long, and 1/2 or 5/8 inch square, it is wedge-shaped on one end to allow insertion fully into the chine, and brass tipped to prevent wear. This rod is calibrated from left to right, beginning at the wedge-shaped tip. The scale is nonlinear, and reads directly in gallons, the graduations being scaled and marked in such a way as to include the conversion from inches to gallons.

The method of using the gauging rod is clearly described in the following passage from a catalogue of coopers' and gaugers' supplies published circa 1922:

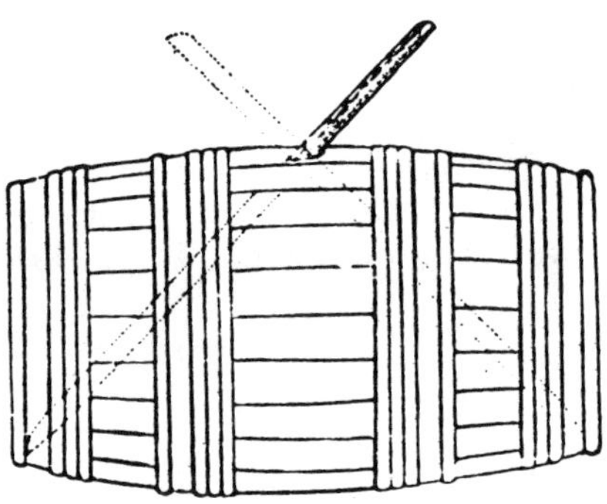

Figure 101: Use of the Gauging Rod

Placing the barrel, for instance, on its side, so as the bung hole will be at the top, insert the sharpened end of the gauge-rod in a slanting direction through the bung-hole until it reaches the place where the head fits into the stave, directly opposite the one in which the bung-hole is, pushing it in as far as it will go. keeping the side on which the number of gallons is marked uppermost. Now, notice exactly where the rod strikes the under side of the middle of the bung-hole, putting your finger on the place if you cannot see it before drawing out the rod. If the bung-hole was exactly in the middle of the barrel every time, that measure would be enough, but as it is not always so, it is necessary to reverse your rod and measure in the same way from the middle of the bung-hole to the other end of the barrel, and take the average. For instance, you measure one way and it shows 49 gallons; you measure the other way and it shows 47 gallons; then you take the average, which is 48 gallons, which is the capacity of the barrel. Be careful not to confound the inch scales on two sides of the rod with the gallon scale.

(Ref. 35)

Once the capacity of the cask is known, then the wantage rod is employed to measure the contents thereof. As in the case of the gauging rod, the wantage rod depends for its accuracy upon the fact that casks are usually made to wine measure. The rod is 16+ Inches long, 5/8 Inch square, and has a narrow brass lip or plate representing the zero point projecting slightly from one side. Wantage Rods are customarily marked with either 8 or 12 scales, each scale being graduated and marked to correspond to a particular nominal capacity cask.

In use, the wantage rod is employed as follows:

With the barrel on its side, bung hole uppermost, insert the Wantage Rod vertically down through the hole until the brass plate is even with the underside of the stave. Then withdraw it, and note where the liquid mark intersects the scale on the Rod appropriate to the cask being measured. The graduation corresponding to that mark indicates the number of Gallons of liquid wanting to fill the cask. *(Ref. 35)*

The Stanley Company manufactured two types of wantage rods in the period after 1855: the No. 37, with 12 scales, and the No. 44, with 8, and two models of gauging rod: the No. 45, for casks up to 120 gallons, and the No. 45½, for casks up to 180 gallons. The No. 45½ was also graduated with wantage tables, thus eliminating the need for a separate wantage rod in most cases.

The Nos. 37 and 45½ had both been discontinued by the time of the merger with The Stanley Works in 1921, and the No. 44 was also dropped shortly thereafter. The No. 45, because of its more general utility, continued to be offered up through 1932.

37 Wantage Rod, 16-1/2 Inch

Offered:	1867 thru 1917
Material:	Maple, Brass Lip
Thickness:	5/8 Inch
Scales:	Wantage: 16 to 200 Gallons, in 12 Lines

38 Carpenters' Caliper Rule, 6 Inch, 2 Fold

Offered:	1855 thru 1922	
Construction:	Square Joint, Left Hand Caliper	
	Leather Case (1912 thru 1917)	
Material:	Ivory, German Silver Hinge, Caliper, & Tips	
Width:	7/8 Inch	
Graduations:	Body:	8ths and 16ths of Inches (1872 and before)
		8ths, 10ths, and 16ths of Inches (1874 thru 1877)
		8ths, 10ths, 12ths, and 16ths of Inches (1879 and after)
	Caliper:	16ths of Inches

Between 1888 (Ref. 14) and 1892 (Ref. 15) the caliper jaw on this rule was changed slightly, to have a rounded back (Fig. 102b) instead of the hollow back (Fig. 102a) which had been customary. No obvious reason can be found for this change; it must have been made for merely cosmetic purposes, or possibly to slightly reduce manufacturing cost (the hollow would have had to be milled, but the rounded back could be satisfactorily shaped by hand on a grinding disk).

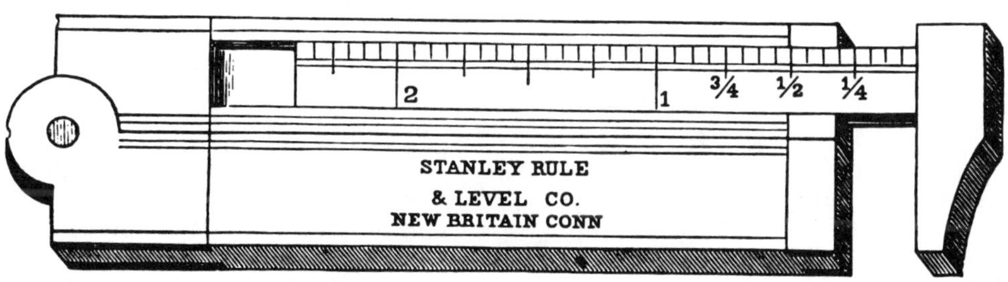

Figure 102a: No. 38, 1867 *(Ref. 7)*

Figure 102b: No. 38, 1892–1898 *(Roger K. Smith Collection)*

This design modification was not confined to the No. 38. The No. 32, the only other Stanley caliper rule for which there exists a pre-1892 catalogue illustration was similarly modified at the same time. Rules wider than 1 inch (such as the Nos. 13 and 36½) were not so modified; it would appear that it had been decided that all of the narrower caliper rules would have a rounded jaw, while the wider would have a hollow. This cannot be known for certain, but examination of catalogue illustrations after this date, combined with study of numerous examples, would seem to support such a hypothesis.

39 Carpenters' Caliper Rule, 1 Foot, 4 Fold

Offered:	1855 thru 1922	
Construction:	Square Joint, Edge Plates, Left Hand Caliper Leather Case (1912 thru 1917)	
Material:	Ivory, German Silver Hinges, Caliper, & Tips	
Width:	7/8 Inch	
Graduations:	Body:	8ths, 10ths, and 16ths of Inches (1872 and before) 8ths, 10ths, 12ths, and 16ths of Inches (1874 and after)
	Caliper:	16ths of Inches

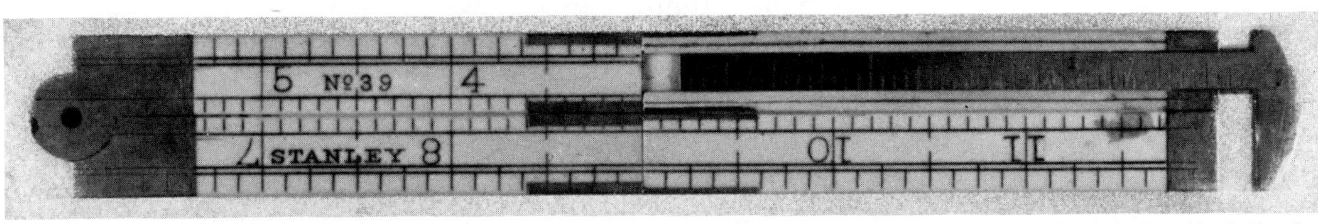

(Eugene Frankio Collection)

Figure 103: No. 39, 1858–1921

40 Carpenters' Caliper Rule, 1 Foot, 4 Fold

Offered:	1855 thru 1922	
Construction:	Square Joint, Full Bound, Left Hand Caliper Leather Case (1912 thru 1917)	
Material:	Ivory, German Silver Hinges, Caliper & Trim	
Width:	1/2 Inch (1860) 5/8 Inch (1862 and after)	
Graduations:	Body:	8ths and 16ths of Inches
	Caliper:	16ths of Inches

It is possible that the No. 40 may have had a caliper graduated in 32nds of inches during the period 1900-1902. According to the catalogues 32nds graduations only became an option in 1905, and standard in 1925 (see SPECIAL ORDER STANDARD RULES, Chapter 4). However, in 1900 and 1902 (Refs. 18 and 20), the catalogue illustration of this rule showed 32nds graduations on the caliper, although the 1898 and before, and 1907 and after, catalogues showed 16ths. It may have had 32nds graduations, but what is more likely is that they used the same line cut to illustrate this rule as the No. 3, the equivalent rule in boxwood. The No. 3 had 32nds graduations, and they were shown in this cut.

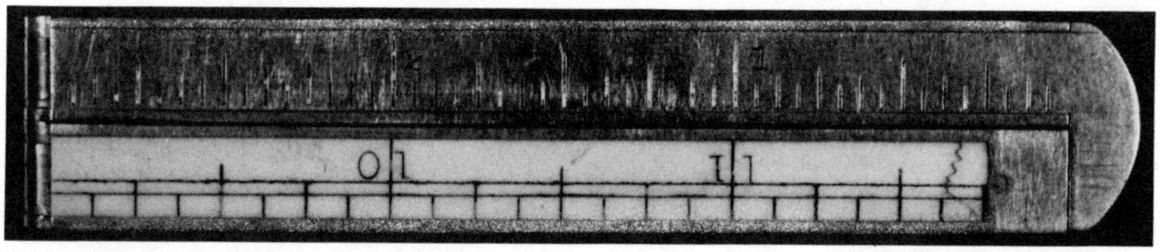

Figure 104: No. 40, 1858–1921 *(Roger K. Smith Collection)*

40½ Carpenters' Caliper Rule, 6 Inch, 2 Fold

Offered: 1888 thru 1922
Construction: Square Joint, Full Bound, Left Hand Caliper
Leather Case (1912 thru 1917)
Material: Ivory, German Silver Hinge, Caliper & Trim
Width: 5/8 Inch
Graduations: Body: 8ths and 16ths of Inches
Caliper: 16ths of Inches

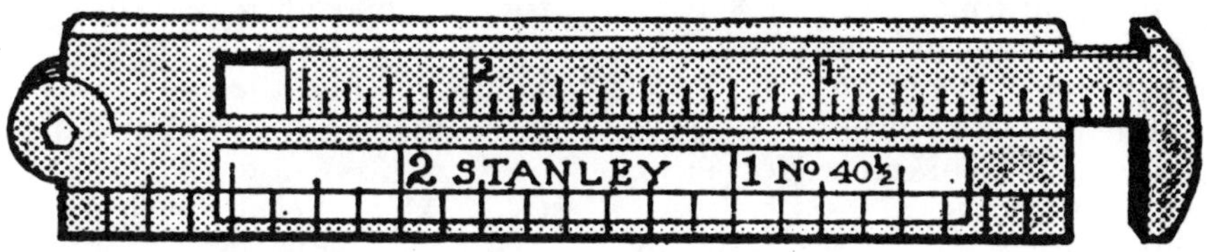

Figure 105: No. 40½, 1914 *(Ref. 29)*

41 Yard Stick, 3 Foot

Offered: 1855 thru 1958
Construction: Polished (1907 and before)
Material: Maple, Brass Tips
Width: 1 Inch
Graduations: 8ths of Inches
Fractions of a Yard

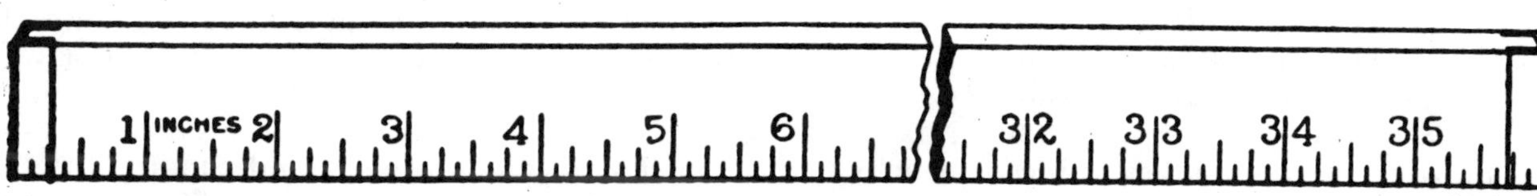

Figure 106: No. 41, 1934 *(Ref. 44)*

41YS Yard Stick, 3 Foot

Offered: 1960 thru 1972

Identical to No. 41; a model number change only

41½ Wood Measure, 5 Foot

Offered: 1855
Construction: Not Known
Width: Not Known
Graduations: Feet and Inches
10ths and 100ths of a Foot

The only information available about the No. 41½ Wood Measure is the name and the bare description above, both derived from the 1855 Price List (Ref. 1). No examples of this rule are known to the author, nor any information which would allow any conjecture as to either appearance or application.

Ship Carpenters' Bevels

The bevel gauge is a widely used woodworkers' tool, consisting (in its most common form) of a thin tongue or blade set into a thick stock, with the joint between the two adjustable, so that the tongue can be set and held at any desired angle to the stock. The purpose of the bevel gauge is to transfer an angle from a drawing or protractor, or from the work itself, to a workpiece for either marking or testing purposes.

Wooden ships are possibly the most extreme example of a structure in which no two pieces join at a right angle. The planks form compound angles as they follow the hull shape, the frames are curved and beveled to fit the planking, and the deck is crowned from side to side and a hollow curve from bow to stern. Because of this the ship carpenter used the bevel gauge more than any other type of woodworker.

The bevel he used was somewhat different from the common Sliding T Bevel used by ordinary carpenters.

In its most common form, it was both longer (12 inches) and slimmer (5/8 inch). It was made of boxwood, and was graduated in 8ths and 16ths of inches along its length, allowing it to double as a rule. It usually had two unequal length tongues, one at either end, attached to the stock with pivoting friction joints, instead of a sliding clamp joint, Variations existed, of course. Sometimes the stock was of rosewood instead of boxwood (in which case the graduations were omitted); sometimes the rule was longer than 12 inches; sometimes it had a single long tongue, instead of two shorter ones.

Ship carpenters' bevels were offered by Stanley from the very beginning. The 1855 Price List of A. Stanley & Co. (Ref. 1) listed both a boxwood (No. 42) and a rosewood (No. 41½) bevel, with the boxwood bevel being available in either a single-tongue or a double-tongue version. After the merger with Hall & Knapp in 1857, the No. 42½ was dropped (Ref. 2), and ten years later, in 1867, the single-tongue boxwood bevel was given its own number (No. 43).

The subsequent history of these two bevels (the Nos. 42 and 43) reflects the decline of the wooden ship building industry. As the number of ship carpenters declined, so did the demand for these rules. The No. 43, being less generally useful, was the first to be discontinued, in 1898 (Ref. 17); the No. 42 continued to be offered until 1922, after which it too was removed from the catalogue (Ref. 34).

The demand for these bevels had not vanished entirely, however, and Stanley briefly reintroduced the No. 42 in 1929 (Ref. 41), only to drop it again the next year in response to worsening economic conditions.

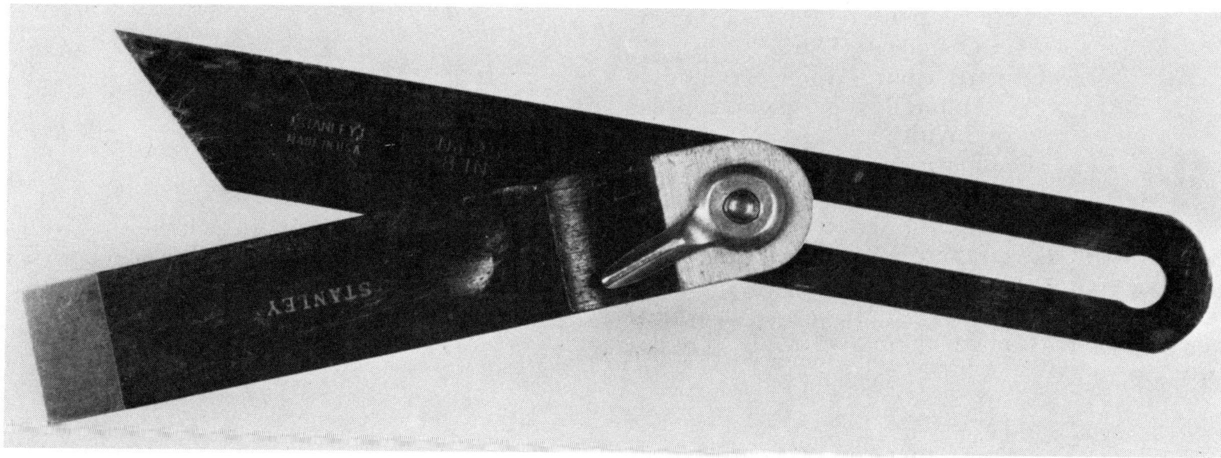

Figure 107: Stanley Sliding T Bevel *(Author's Collection)*

42 Ship Carpenters' Bevel, 1 Foot

Offered:	1855 thru 1922, 1929
Construction:	Single or Double Tongue (1862 and before) Double Tongue (1865 and after)
Material:	Boxwood, Brass Tongue(s) & Trim
Width:	5/8 Inch
Graduations:	8ths and 16ths of Inches

As originally listed in 1855, the No. 42 was made in two versions, one with two short tongues, one at each end, the other with a single long tongue. In 1865 (Ref. 6) this second version was given its own number (No. 43), and thereafter the No. 42 was only available in the double-tongue version.

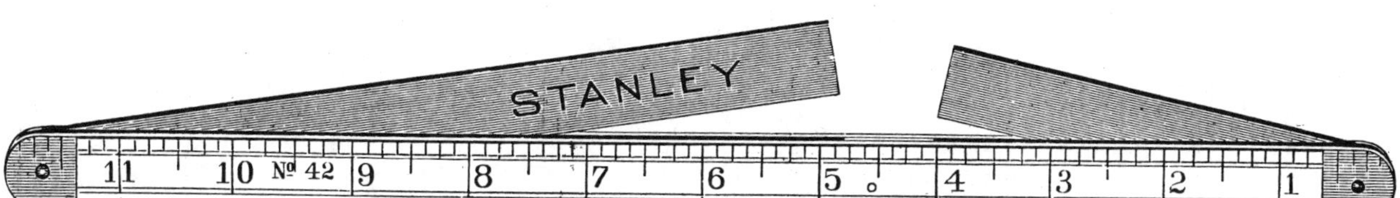

Figure 108: No. 42, 1898 *(Ref. 17)*

The No. 42 ship carpenters' bevel appeared in every catalogue up to and including 1922, and also in two catalogues thereafter, the 1929 dealers' catalogue (Ref. 41), and the 1942 corrected price list (Ref. 52). It has been assumed that the 1929 listing was a genuine attempt to resume production, canceled a year later at the advent of the depression of 1929–1939, but that the second reappearance in 1942 was simply an effort to use up existing stocks.

42½ Ship Carpenters' Bevel, 14 Inch

Offered:	1855
Construction:	Single Tongue
Material:	Rosewood, Brass Tongue & Trim
Width:	¾ Inch
Graduations:	None

Tools made of exotic woods such as rosewood, live oak, etc., are more often found in the kits of ship carpenters than any other group of woodworkers. There was a tendency among them to make their own tools, due to the rather specialized work which they did, and since most of them worked near seaports, they had access to woods which were denied the ordinary cabinetmaker.

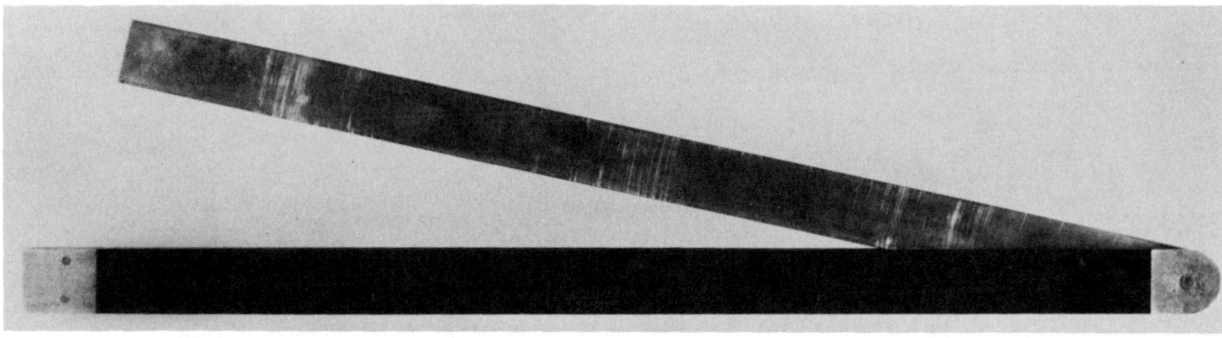

Figure 109a: No. 42½, 1855 *(Author's Collection)*

It is not surprising, therefore, that Stanley should manufacture a premium bevel of rosewood to appeal to ship carpenters whose planes, tool handles, etc., were extra fancy. The No. 42½, larger than the No. 42, was appealing to the eye (and probably to the hand as well), and would be perfectly at home in the tool chest of some "Chips."

This bevel was only offered for a short time, however, being only listed in the 1855 price list (Ref. 1). Apparently its quality and appearance could not offset its premium price (50 percent more than the 12 Inch No. 42), and there was not sufficient demand to warrant continued production*.

Although no longer listed in the catalogue, the No. 42½ was available as a special order item for at least the next sixty to seventy years. Examples have been noted bearing trademarks from the period 1900-1910, and from the period 1921-1936.

At some time after 1877 the No. 42½ was modified to use the lever-type clamping joint (U.S. Patent #194,851) invented by Justus A. Traut, which was introduced on Stanley Sliding T Bevels at that date. Apparently the friction joint used on the No. 42½ could not be made stiff enough to properly hold the 14 inch blade, and some stronger method was required.

*: Properly speaking, the No. 42½, having no graduations, was not a rule, and should not have been included in this book. However, since the other Ship Carpenters' Bevels, the Nos. 42 and 43, were described, it was decided that the No. 42½ would be also, to give at least the illusion of uniformity.

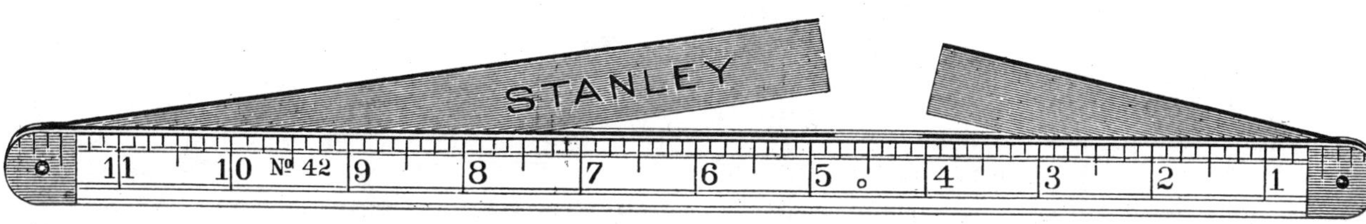

Figure 109b: No. 42½, Ca. 1900–1920 *(Charles and Walter Jacob Collection)*

43 Board Stick, 3 Foot

Offered:	1859
Construction:	Flat
Material:	Brass Head
Width:	Not Known
Scales:	Board Measure

43 Ship Carpenters' Bevel, 1 Foot

Offered:	1865 thru 1892
Construction:	Single Tongue
Material:	Boxwood, Brass Tongue & Trim
Width:	5/8 Inch
Graduations:	8ths and 16ths of Inches

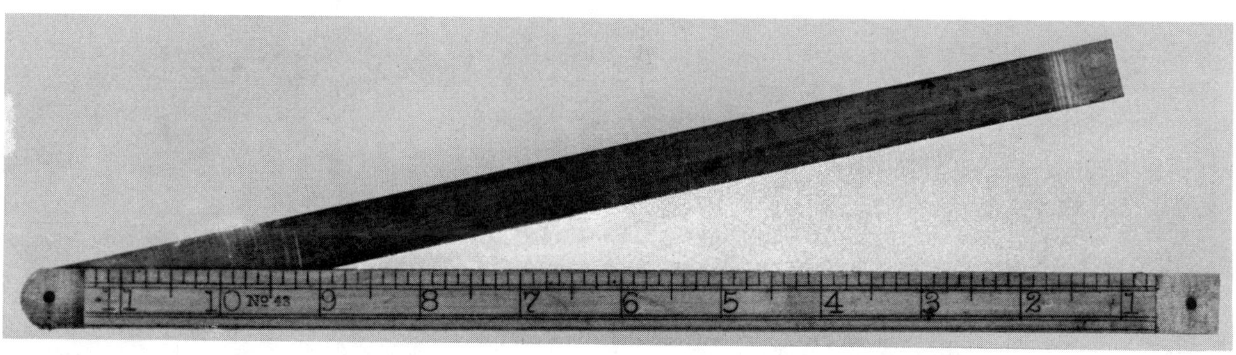

Figure 110: No. 43, 1865–1892 *(Author's Collection)*

43½ Board Stick, 3 Foot

Offered: 1859 thru 1915
Construction: Flat, Cast Oval Head
Material: Hickory, Brass Head & Tip
Width: 1 Inch
Scales: Board Measure: 12 to 22 Feet, in 6 Lines

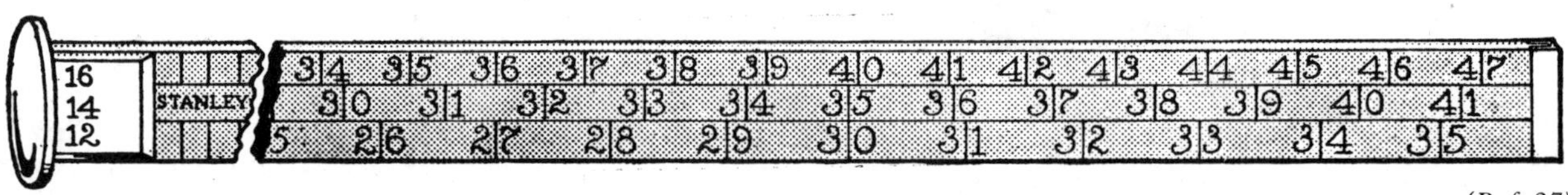

Figure 111: No. 43½, 1912 *(Ref. 27)*

44 Board Stick, 2 Foot

Offered: 1855
Construction: Square
Width: Not Known
Scales: Board Measure

44 Wantage Rod, 16½ Inch

Offered: 1859 thru 1922
Construction: Square
Material: Maple, Brass Lip
Width: ⅝ Inch
Scales: Wantage: 16 to 120 Gallons, in 8 Lines (1898 and before)
16 to 48 Gallons, in 8 Lines (1909 and after)

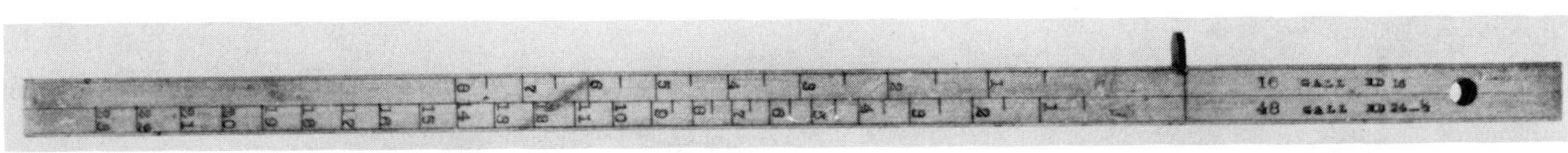

Figure 112: No. 44, 1859–1922 *(Charles and Walter Jacob Collection)*

45 Board Stick, 3 Foot

Offered: 1855
Construction: Square
Width: Not Known
Graduations: Not Known
Scales: Board Measure

45 Gauging Rod, 3 Foot

Offered: 1859 thru 1932
Construction: Square, One End Wedge Shaped
Material: Maple, Brass Tip
Width: ½ Inch
Graduations: 10ths of Inches
Scales: Gauging: 120 Gallons
Wantage for 42 Gallon Cask (1925 and after)

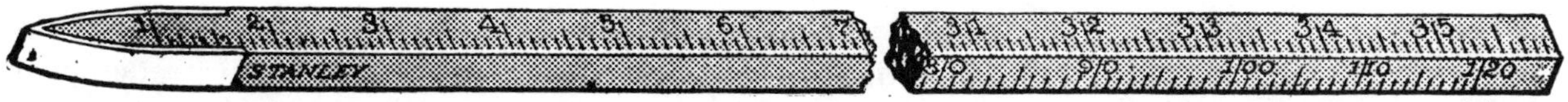

Figure 113: No. 45, 1915 *(Ref. 31)*

45½ Gauging Rod, 4 Foot
Offered: 1862 thru 1915
Material: Maple, Brass Tip
Width: ⅝ Inch
Scales: Gauging: 180 Gallons
Other Markings: Wantage Tables

46 Board Stick, 2 Foot
Offered: 1855 thru 1915
Construction: Octagonal
Material: Maple, Brass Cap and End Plate
Thickness: ⅞ Inch
Scales: Board Measure: 8 to 23 Feet, in 16 Lines

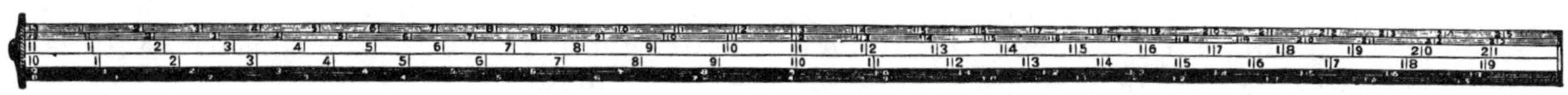

Figure 114: No. 46, 1872 *(Ref. 9)*

46½ Board Stick, 2 Foot
Offered: 1870 thru 1915
Construction: Square
Material: Maple, Brass Cap and End Plate
Thickness: ⅞ Inch
Scales: Board Measure: 8 to 23 Feet, in 16 Lines

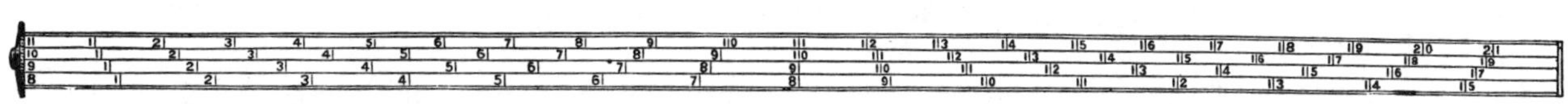

Figure 115: No. 46½, 1888 *(Ref. 14)*

47 Board Stick, 3 Foot
Offered: 1855 thru 1915
Construction: Octagonal
Material: Maple, Brass Cap and End Plate
Thickness: ⅞ Inch
Scales: Board Measure: 8 to 23 Feet, in 16 Lines

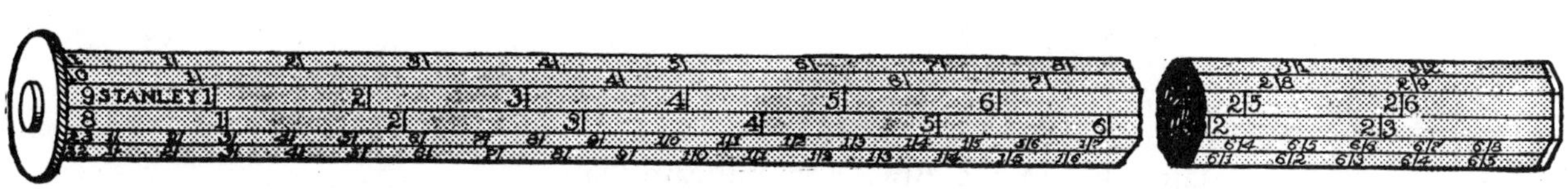

Figure 116: No. 47, 1915 *(Ref. 31)*

Board & Log Canes

It is not known whether the idea of a 3 foot measuring stick shaped like a cane, with a brass head and ferrule, was original with A. Stanley & Co., or had already been invented prior to the company's founding. Certainly such canes were part of their product line from the first.

Initially the only such cane offered (Ref. 1) was the No. 48 Board Cane, marked with board scales, and intended to convert the dimensions of sawn lumber into board measure. A second cane was added, within four years (Ref. 2), the No. 48½ Log Cane, marked with log scales, for use in estimating the amount of sawn lumber which could be realized from unsawn logs. Six years later, in 1865 (Ref. 6), a third cane, the No. 47½, was added to the product line, another board cane, similar to the No. 48, but more cheaply made.

The combination of cane and measuring stick is an interesting one, and conjures up the image of an elderly sawyer or timber buyer, clambering over piles of logs and sawn lumber with the aid of his cane, and then laying it across a log or board and recording the indicated figure.

The board and log canes were designed for such rough usage. They were made of hickory (except in the case of the No. 47½), an extremely tough wood, and were tipped with a brass ferrule to reduce wear. The round brass head (cast, except for the No. 47½) served a double duty, giving the hand a place to grip when used as a cane, and also serving as a shallow hook to aid in placing the head of the cane at the far side of the board/log. The cane itself was octagonal, with a single scale marked on each of the eight faces, and was tapered from ¾ inch thick at the head to ½ inch at the ferrule.

The No. 47½ was only offered for a very short time, probably due to poor acceptance by the public because of its lower quality. The other two canes fared better; both were continued in production for over fifty years, and only discontinued by Stanley after 1915 (Ref. 31), at the same time that all other rules with board tables or board scales were also discontinued.

47½ Board Cane, 3 Foot

Offered:	1865
Construction:	Octagonal, Tapered
Material:	Wood, Brass Head & Tip
Thickness:	¾ Inch at Head
Scales:	Board Measure, 9 to 16 feet, in 8 lines

The No. 47½ board cane was apparently an attempt to market a less expensive version of the No. 48 board cane which had been offered since 1855. Presumably, since the wood used to make this cane was not identified, it was not hickory, but some cheaper substitute, such as maple. Similarly, the head was described in the price list (Ref. 2.8) as "brass", instead of "cast brass," as in the case of the No. 48; this difference in description must infer some difference in strength or quality between the two styles of head as well. The price difference between the two canes was only 20 percent ($15.00 vs. $18.00 per dozen). This probably accounts for the fact that the No. 47½ was only offered in 1865; the saving of 33 cents in the cost of the cane, even in the currency of that time, was not enough to offset the lower quality of this cane relative to the No. 48.

47½ Board Stick, 3 Foot

Offered:	1870 thru 1915
Construction:	Square
Material:	Maple, Brass Cap and End Plate
Thickness:	7/8 Inch
Scales:	Board Measure: 8 to 23 Feet, in 16 Lines

48 Board Cane, 3 Foot

Offered: 1855 thru 1915
Construction: Octagonal, Tapered, Cast Head and Tip
Material: Hickory, Brass Head & Tip
Thickness: ¾ Inch at Head
Scales: Board Measure: 9 to 16 Feet, in 8 Lines

Figure 117: No. 48, 1898

Log Rules and Log Measures

One of the most persistent problems which faces the buyer or seller of saw logs is that of estimating the amount of sawn lumber which a log will yield when taken to the mill.

Over the years many efforts have been made to develop formulas, tables, and scales that would perform this estimate, given the diameter and length, and yield as a result the number of board feet which a sawmill could expect from a log.

The table or formula for performing such an estimate is called a "log rule," and devising such a rule is a very complex problem, involving a large number of variables and near-intangible factors. Allowance must be made for the shape of the log, with taper and sweep taken into consideration. Wastage during the sawing process must also be considered, and expected losses due to saw kerfs, slabs, edgings, and shrinkage must be accounted for. Finally, account must also be taken of the types and sizes of timbers and boards being cut by the mill in question.

Over the years a great many attempts have been made to develop good log rules, some intended for general use, others to deal with specific local conditions or specific types of wood. As a result, the number of different rules has proliferated anarchistically, and today in the United States and Canada alone there are more than 95 different recognized rules, many known by more than one name!

Three methods exist for developing such a rule. The first is to use actual measurements, recording over a period of time the diameter of each log as it enters the mill and the amount of sawn lumber which results from it. The statistical results of these records is called a "mill tally" log rule, and is very accurate, as long as the conditions under which the original data were obtained prevail.

The second method is to draw diagrams on various size circles representing typical sawing patterns, with allowance made for saw kerfs and shrinkage. All wastage due to slabs and edgings can be identified from such diagrams, and the resulting number of board feet per foot of log calculated. Rules developed in this manner are known as "diagram" log rules, and will be accurate or inaccurate depending upon the extent to which real conditions correspond to the allowances used.

The third method is to develop a mathematical formula which corresponds fairly closely to actual sawmill results (sort of fitting a formula to the results of the measurements in method 1), and then using the formula as the rule. This type of rule is called a "formula" log rule, and is applicable or not depending on how good the original statistics were, and how carefully the original developer fitted his equation to them.

Two of the most commonly used log rules during the nineteenth century were those of Edward Doyle, published in 1825, and J.M. Scribner, published in 1846.

The Doyle rule is a formula rule, with the formula usually written in the form:

$$BF = (D-4)^2 \times L/16$$

Where:

D = Diameter of Log in Inches
L = Length of Log in Feet
BF = Expected Sawn Lumber Yield in Board Feet

This rule was notoriously inaccurate (in the favor of the buyer) when scaling small logs, giving an error of nearly 50 percent, for example, for logs 10 inches in diameter. Because of (or perhaps in spite of) this inaccuracy, the Doyle rule is one of the most extensively used log rules in the United States and Canada, particularly in mills dealing in small logs. Freese (Ref. 131) characterizes it as "one of the most widely used and roundly cursed log rules in existence."

The Scribner rule is a diagram rule, developed in 1846 by J. M. Scribner, a country clergyman, and published in a pocket-sized book of useful interest, wage, board measure, etc., tables entitled *The Ready Reckoner; for Ship Builders, Boat Builders, and Lumber Merchants.* Scribner's rule is fairly accurate and consistant for logs smaller than 16 feet in length and 28 inches in diameter, but tends to slightly favor the buyer for timber larger than this.

The Doyle rule was rapidly replaced by the more accurate Scribner rule, going out of print in the 1850's, and would be little used today, except for a rather strange series of events which revived it. In 1876 a publisher, George W. Fisher, who had recently purchased the rights to Scribner's *Ready Reckoner,* decided to reissue it. This same publisher had also previously acquired the rights to Doyle's *Ready Reckoner,* and for some unknown reason substituted Doyle's tables (with some minor revisions) for Scribner's in the new edition, *Scribner's Lumber and Log Book.*

As a result of this republication, Doyle's rule received a new lease on life, and under various alternate names, such as "Doyle's Revised" rule, "New Scribner" rule, and "New" rule, was in common use for many years thereafter.

48½ Log Cane, 3 Foot

Offered:	1859 thru 1915
Construction:	Octagonal, Tapered, Cast Head and Tip
Material:	Hickory, Brass Head & Tip
Thickness:	¾ Inch at Head
Graduations:	8ths of Inches
Scales:	Log Measure, 12 to 22 Feet, in 7 or 8 Lines
	Rule: Not Known (1872 and before)
	Scribners' (1874 thru 1877)
	"Doyle's Revised" (1879 and after)

The No. 48½ Log Cane was identical in appearance and construction to the No. 48 Board Cane, differing from it only in the type of graduations. Where the No. 48 was marked in board scales, the No. 48½ was marked instead with log scales (see LOG RULES AND LOG MEASURES, in this chapter).

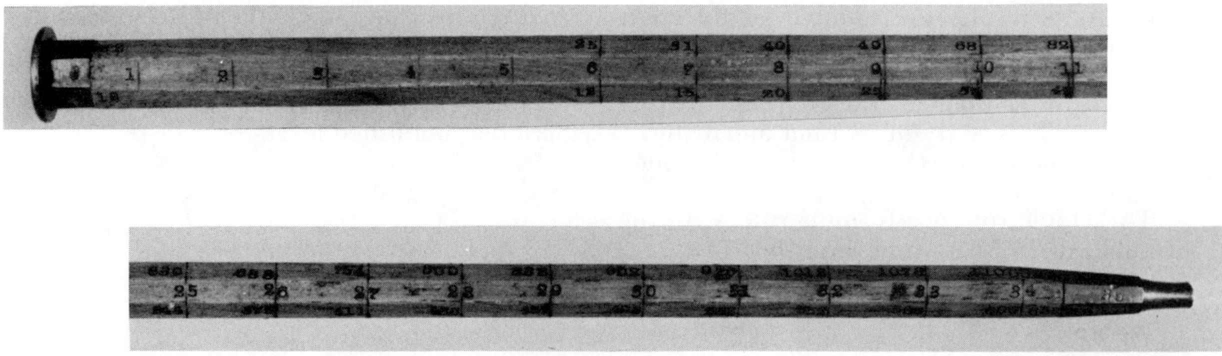

Figure 118: No. 48½ (Scribner's Scales), 1859–1877 *(Paul Kebabian Collection)*

It is not known which log rule initially served as a basis for the scales on the No. 48½. The catalogues from 1859 (Ref. 2) through 1872 (Ref. 9) do not identify it, and the best that can be done is to surmise, based upon the acceptance being accorded various log rules at that time, that it was probably Scribner's (this would be consistent with the fact that the 1874 (Ref. 10) and 1877 (Ref. 11) catalogues identify Scribner's rule by name). All known examples of the No. 48½ are marked either with this rule, or with Doyle's (see below).

In 1879, the scales on the No. 48½ were redesigned to base them upon the newly-revived "Doyle's Revised" rule. Apparently the log purchasers were the primary buyers of these log canes, and thus the demand was for a set of log scales which would be advantageous to the mill.

The scales on these log canes were arranged differently, depending on whether they were marked by Scribner's or Doyle's rule. On those marked by Scribner's rule, only seven of the eight faces were marked with log scales. The eighth was graduated in inches, beginning at the head and extending to the ferrule. The seven log scales, for logs 12, 13, 14, 16, 18, 20, and 22 feet long, began 6 inches from the head and extended to the ferrule.

Canes marked by Doyle's rule had eight log scales, one on each face, all beginning 12 inches from the head. These were scales for logs 12, 14, 15, 16, 17, 18, 19, and 20 feet long. The face with the 12 foot scale was graduated in inches from the head to the start of the scale.

49 Drafting Scale, 1 Foot

Offered:	1855
Construction:	Not Known
Material:	Boxwood
Width:	Not Known
Graduations:	Not Known

49 Board Stick, 3 Foot

Offered:	1859 thru 1915
Construction:	Flat, "Extra Strong"
Material:	Hickory, Brazed Steel Cap, Brass Tip
Width:	1 Inch
Scales:	Board Measure: 12 to 22 Feet, in 6 Lines

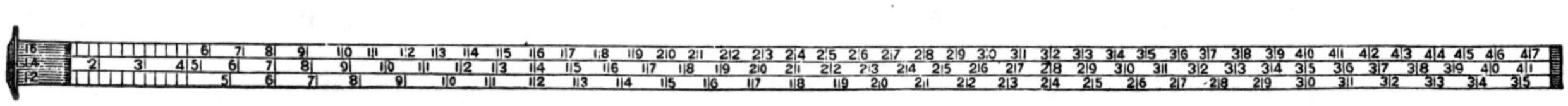

Figure 119: No. 49, 1879 *(Ref. 12)*

Forwarding Stick(s)

It is possible that "the" No. 49½ forwarding stick was actually two different sticks, the first offered from 1862 to 1877, and the second from 1879 to 1892.

The basis for this supposition is as follows: In 1862 and 1865 (Refs. 5 and 6) the No. 49½ was listed by itself in a separate section of the catalogue, "Forwarding Sticks"; from 1867 to 1877 (Refs. 7 through 11) it was included in the section, "Wantage and Gauging Rods"; then from 1879 through 1892 (Refs. 12 through 15) in the section "Board, Log, and Wood Measures," and was described as having a "cast brass (T) head." It is not reasonable that any single measuring stick could be classified as a "Gauging or Wantage Rod" at one time, and at another as a "Board, Log, or Wood Measure." Additionally, a (T) head, such as was ascribed to the stick from 1879 on would be a natural adjunct to a wood measure, but useless on a gauging rod.

The most likely interpretation of these facts is that Stanley manufactured two No. 49½ forwarding sticks. The first, offered from 1862 through 1877 (assuming that the stick of 1862 and 1865 was the same as that of 1867 through 1877) had no (T) head, and was intended for some type of liquid or wine measure. The second, offered from 1879 through 1892, had some for of (T) head, and was intended for use by forwarding (freight) agents. The two sticks may or may not have had the same scales, or the same cross-sectional shapes, etc.

49½ Forwarding Stick, 5 Foot
Offered: 1862 thru 1877
Construction: Not Known
Material: Brass Head
Width: Not Known
Graduations: Not Known

49½ Forwarding Stick, 5 Foot
Offered: 1879 thru 1892
Construction: Cast "T" Head
Material: Brass Head
Width: Not Known
Graduations: Not Known

50 "Scholars'" Rule, 1 Foot
Offered: 1855 thru 1860
Construction: Not Known
Material: Boxwood
Width: Not Known
Graduations: Not Known

50 Yard Stick, 3 Foot
Offered: 1862 thru 1942
Construction: Polished
Material: Hickory, Brass Tips
Width: ¾ Inch
Graduations: 8ths of Inches
Fractions of a Yard

In 1943 the No. 50, along with a few others, was added to the list of rules canceled to conserve materials and to make room for increased war production. The price list did not identify this cancellation as either "permanent" or "temporary," but it must have been the former; when the first postwar catalogue was published in 1947 (Ref. 54) the No. 50 was no longer part of the product line.

Figure 120: No. 50, 1922–1932 *(Author's Collection)*

51 Carpenters' Rule, 2 Foot, 4 Fold
Offered: 1855 thru 1941
Construction: Arch Joint, Middle Plates
Material: Boxwood, Brass Hinges & Tips
Width: 1 Inch
Graduations: 8ths and 16ths of Inches (1872 and before)
8ths, 10ths, and 16ths of Inches (1874 thru 1877)
8ths, 10ths, 12ths, and 16ths of Inches (1879 and after)
Scales: Drafting

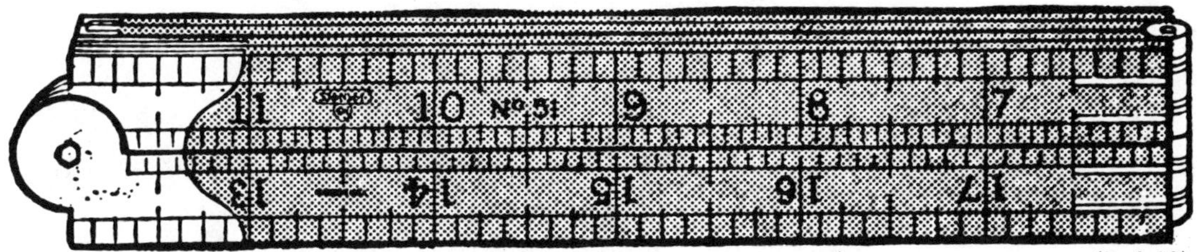

Figure 121: No. 51, 1926 *(Ref. 39)*

52 Carpenters' Rule, 2 Foot, 4 Fold

Offered: 1859 thru 1917
Construction: Arch Joint, Half Bound
Material: Boxwood, Brass Hinges & Trim
Width: 1 Inch
Graduations: 8ths and 16ths of Inches (1872 and before)
8ths, 10ths, and 16ths of Inches (1874 thru 1877)
8ths, 10ths, 12ths, and 16ths of Inches (1879 and after)
Scales: Drafting

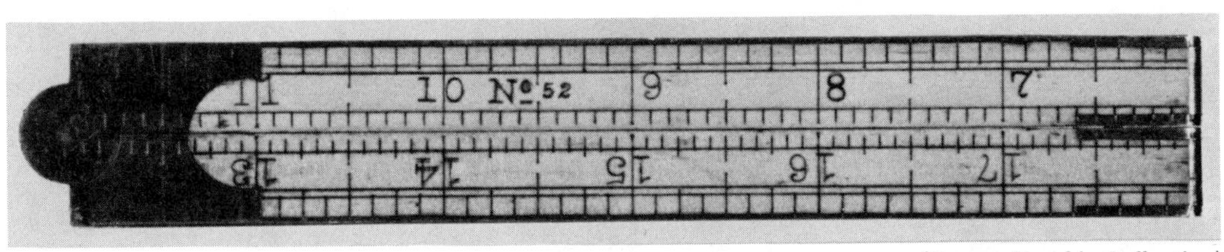

Figure 122: No. 52, 1859—1898 *(Eugene Frankio Collection)*

53 Carpenters' Rule, 2 Foot, 4 Fold

Offered: 1855 thru 1934
Construction: Arch Joint, Edge Plates
10ths, Scale on Edge (1860 and before)
10ths Scale on Surface (1919 and after)
Material: Boxwood, Brass Hinges & Tips
Width: 1 Inch
Graduations: 8ths, 10ths, and 16ths of Inches (1877 and before)
8ths, 10ths, 12ths, and 16ths of Inches (1879 and after)
Scales: Drafting

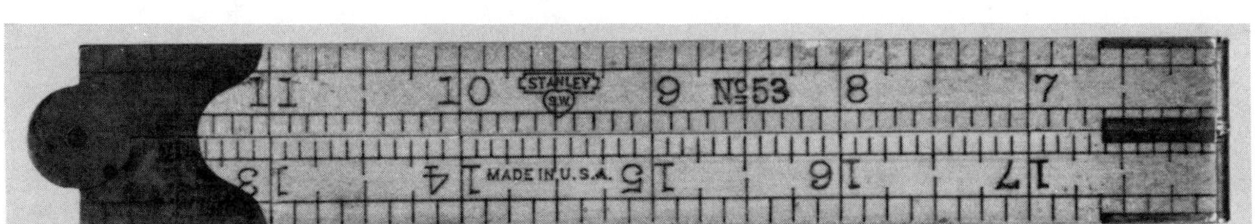

Figure 123: No. 53, 1922—1932 *(Author's Collection)*

Architects' Folding Rules

The inside edges (the edges brought together by closing the rule joint) of the Nos. 53½ and 86½ architects' rules were beveled on their inside surfaces to bring the scales in those locations closer to the surface when the rule was lying flat. This was only done for the central 4 inches of each leg; the last inch at each end was left unbeveled in order to allow the joints and tips to be fitted as on a conventional rule (see Fig. 124).

Figure 124: Joint and Bevel Spacing on an Architects' Rule *(Author's Collection)*

The scales marked on these beveled surfaces were drafting scales (1/8 , 1/4 , 3/8 , and 1/2 inch per foot) representing scale distances of 29' 12", 14' 12", 9' 12", and 7' 12" respectively. Bringing these scales closer to the surface being scaled greatly improved the accuracy when either laying out work or taking off dimensions.

53½ Carpenters'/Architects' Rule, 2 Foot, 4 Fold

Offered:	1879 thru 1942
Construction:	Arch Joint, Edge Plates, Inside Beveled Edges
Material:	Boxwood, Brass Hinges & Tips
Width:	1 Inch
Graduations:	8ths, 10ths, 12ths, and 16ths of Inches
Scales:	Drafting

In 1943 the No. 53½, along with a few others, was added to the list of rules canceled to conserve materials and to make room for increased war production. The price list did not identify this cancellation as either "permanant" or "temporary", but it must have been the former; when the first postwar catalogue was published in 1947 (Ref. 54) the No. 53½ was no longer part of the product line.

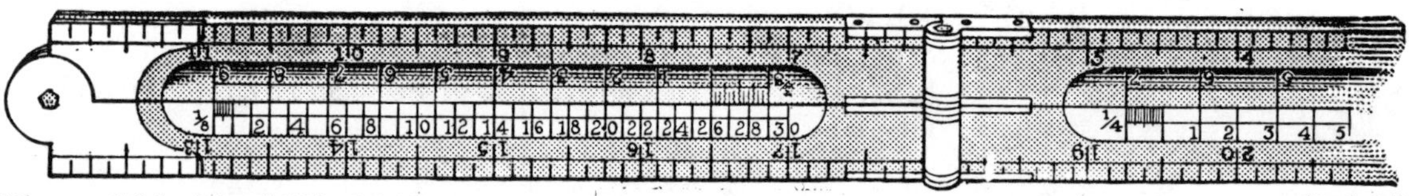

Figure 125: No. 53½, 1915 *(Ref. 31)*

54 Carpenters' Rule, 2 Foot, 4 Fold

Offered: 1855 thru 1932
Construction: Arch Joint, Full Bound
Material: Boxwood, Brass Hinges & Trim
Width: 1 Inch
Graduations: 8ths, and 16ths of Inches (1865 and before)
8ths, 10ths, and 16ths of Inches (1867 thru 1870)
8ths and 16ths of Inches (1872)
8ths, 10ths, and 16ths of Inches (1874 thru 1877)
8ths, 10ths, 12ths, and 16ths of Inches (1879 and after)
Scales: Drafting

It is not clear whether the 10ths scale was added to and deleted from the No. 54 rule as often as the catalogues would have us believe, or whether this is simply a misprint in one or more of those catalogues. The latter case would seem the more likely, in view of Stanley's normal rate of changing the product during this period. If so, then the misprint probably occurred in the 1872 catalogue (Ref. 9), and the 10ths scale was a feature of the No. 54 for the entire period from 1867 through 1877.

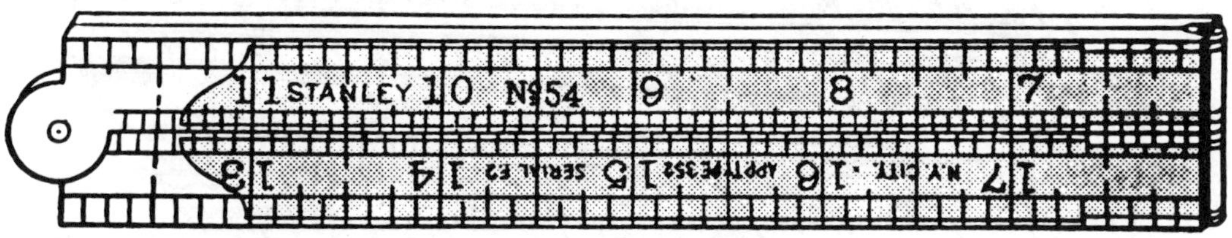

Figure 126: No. 54, 1919 *(Ref. 33)*

Optional Width

In 1860 (Ref. 3) all but two of the 1 foot, 4 fold rules were offered in a choice of widths, 1/2 or 5/8 inch. Those two, the No. 32 (boxwood, with caliper) and the No. 91 (ivory), were both broad rules (1 inch and 7/8 inch wide respectively), and thus were not included in this offer.

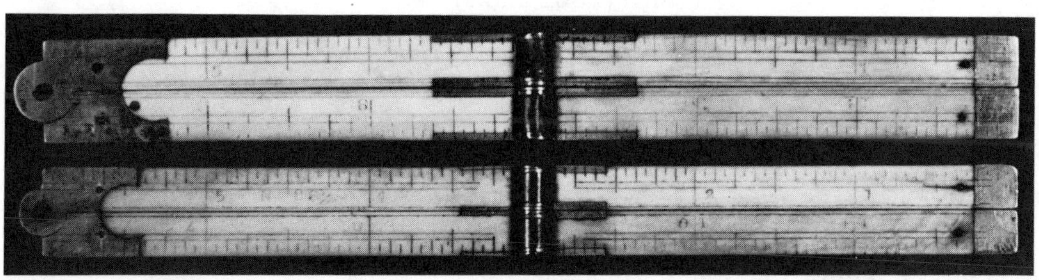

Figure 127: Normal and "Narrow" No. 88½ Rules *(William Baader Collection)*

This was the only time in Stanley history that optional width was so generally available. Previous catalogues (Refs. 1 and 2) had only offered optional width on one of these rules (the No. 90), the other eleven being either "narrow" (the Nos. 92 and 92½), or normal width (the Nos. 55, 56, 57, 64, 65, 65½, 69, 88, and 88½).

Within two years this choice of widths was withdrawn. In the 1862 catalogue (Ref. 5) these twelve rules were once again only available in a single width, 5/8 inch (except for the No. 90; for some reason Stanley was unwilling to give the width of this rule. It was not until 1909 (Ref. 25) that this dimension of the No. 90 was finally listed in the catalogue).

In the process of offering and then not offering this width option the Nos. 92 and 92½ had become 5/8 inch rules like the others, and 1/2 inch rules were no longer available (unless the unspecified width of the No. 90 was, in fact, ½ inch). In a curious twist Stanley at this time elected to bring their rule nomenclature in line with that adopted in 1859 by the Convention of Rule Manufacturers (Editors notes, Ref. 4), with the result that the term "narrow", which had hitherto meant 1/2 inch rules, was now applied to 5/8 inch rules instead.

55 Carpenters' Rule, 1 Foot, 4 Fold

Offered:	1855 thru 1915
Construction:	Arch Joint, Middle Plates
Material:	Boxwood, Brass Hinges & Tips
Width:	1/2 or 5/8 Inch (1860 and before)
	5/8 Inch (1862 and after)
Graduations:	8ths and 16ths of Inches

Figure 128: No. 55, 1859 *(Ref. 2)*

56 Carpenters' Rule, 1 Foot, 4 Fold

Offered:	1855 thru 1915
Construction:	Arch Joint, Edge Plates
Material:	Boxwood, Brass Hinges & Tips
Width:	1/2 or 5/8 Inch (1860 and before)
	5/8 Inch (1862 and after)
Graduations:	8ths and 16ths of Inches

Figure 129: No. 56, 1855–1915 *(William Baader Collection)*

57 Carpenters' Rule, 1 Foot, 4 Fold

Offered:	1855 thru 1915
Construction:	Arch Joint, Full Bound
Material:	Boxwood, Brass or German Silver Hinges & Trim (1859 and before) Boxwood, Brass Hinges & Trim (1860 and after)
Width:	1/2 or 5/8 Inch (1860 and before) 5/8 Inch (1862 and after)
Graduations:	8ths and 16ths of Inches

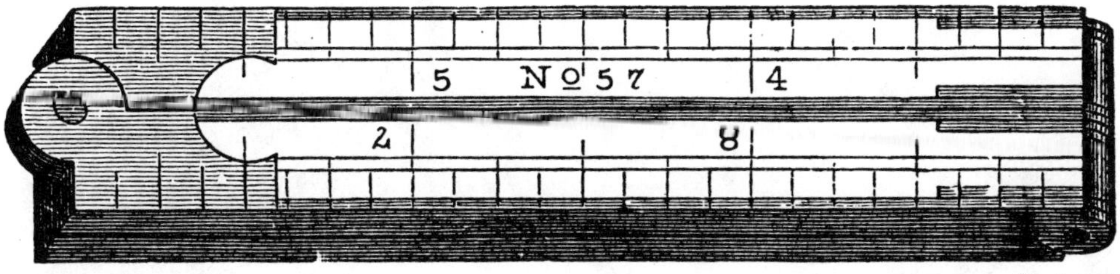

Figure 130: No. 57, 1877 *(Ref. 11)*

This is the only Stanley boxwood rule that was ever made with German silver trim. For a short period, up until 1859 (Ref. 2), the No. 57 was offered with a choice of brass or German silver trim. This was probably a rule pattern left over from either Seth Savage or Seymour & Churchill, and was discontinued as soon as existing stocks were used up. After 1860 (Ref. 4) this choice was no longer offered.

The Six Fold Rule

The 2 foot, 6 fold rule, measuring only four inches when folded, was very easy to carry in the pocket, and had much less tendency to drop out when the workman leaned over. To offset this, however, were the number and location of the joints. The 6 Fold rule had knuckle joints protruding on both sides, as compared with the 4 Fold rule, which had them on one side only; thus there was no way it could be laid flat (except on edge) while taking a measurement. Additionally, as pointed out in Crussel (Ref. 117), many carpenters, instead of using the figures on the rule, calculated the distance by counting inches from the nearest joint, and it must have been very confusing for such a worker to use a 6 Fold rule, where the joints were not at 6, 12, and 18 inches, but at 4, 8, 12, 16, and 20 inches instead.

Stanley made and sold three different types of 6 fold rules. The first, the No. 58, was introduced in 1859, and the second, the No. 58½, a year later in 1860. These rules were identical in construction, differing only in their markings; the No. 58 was marked with 8ths and 16ths on both sides, while on the No. 58½ the scales on one side were replaced with tables of the weights of the common sizes of various materials. The No. 58½ was dropped after only 3 years (Ref. 6), probably because of cost; at $24.00 per dozen its price was nearly that of the fullbound No. 16 engineers' rule.

In 1879 (Ref. 12) a third 6 fold rule, also called the No. 58½, was introduced. This rule was identical to the No. 58, except for being full bound instead of having edge plates, and was graduated with the same scales. This second No. 58½ rule was the last addition to the Stanley offering of 6 Fold rules. The Nos. 58 and 58½ remained in production until 1915 (Ref. 31), when, like so many others, they were dropped to make room for war-related products. They were never returned to production after the end of the war.

The 6 fold rule, requiring 5 joints, was a much more labor-intensive product than the 4 fold rule, which needed only 3. This higher manufacturing cost, coupled with the smaller demand for this type of rule, caused the price of a 6 fold rule to be markedly higher than a 4 fold rule of similar quality (for example: In 1879 [Ref. 12] the No. 58 sold for $13.00 per dozen, while the 4 Fold No. 53 was priced at only $8.00).

58 Carpenters' Rule, 2 Foot, 6 Fold

Offered:	1859 thru 1915
Construction:	Arch Joint, Edge Plates
Material:	Boxwood, Brass Hinges & Tips
Width:	¾ Inch
Graduations:	8ths and 16ths of Inches (1877 and before) 8ths, 10ths, 12ths, and 16ths of Inches (1879 and after)

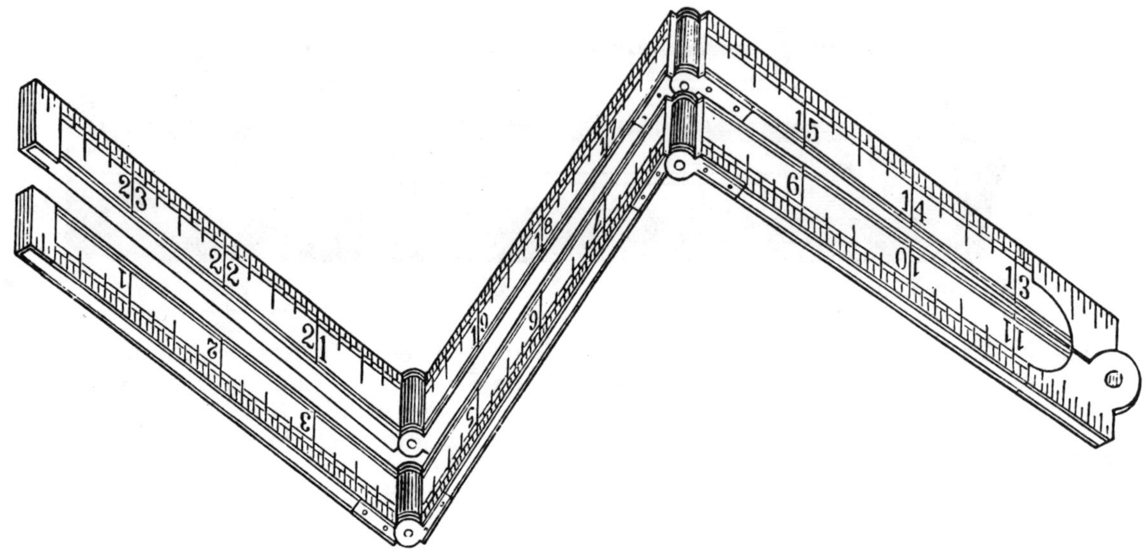

Figure 131: No. 58, 1867 *(Ref. 7)*

58½ Carpenters' Rule, 2 Foot, 6 Fold

Offered:	1860 thru 1862
Construction:	Arch Joint, Edge Plates
Material:	Boxwood, Brass Hinges & Tips
Width:	¾ Inch
Graduations:	8ths and 16ths of Inches
Other Markings:	"Tables for ascertaining the Weights of all sizes of Iron, Steel, Copper, Brass, Lead, etc."

58½ Carpenters' Rule, 2 Foot, 6 Fold

Offered:	1879 thru 1915
Construction:	Arch Joint, Full Bound
Material:	Boxwood, Brass Hinges & Trim
Width:	¾ Inch
Graduations:	8ths, 10ths, 12ths, and 16ths of Inches

Figure 132: No. 58½, 1879–1915 *(William Baader Collection)*

59 Carpenters' Rule, 2 Foot, 4 Fold

Offered: 1855 thru 1915
Construction: Double Arch Joint, Middle Plates, Bitted (1909 and before)
Double Arch Joint, Edge Plates (1912 and after)
Material: Boxwood, Brass Hinges & Trim
Width: 1 Inch
Graduations: 8ths and 16ths of Inches (1872 and before)
8ths, 10ths, and 16ths of Inches (1874 thru 1877)
8ths, 10ths, 12ths, and 16ths of Inches (1879 and after)
Scales: Drafting

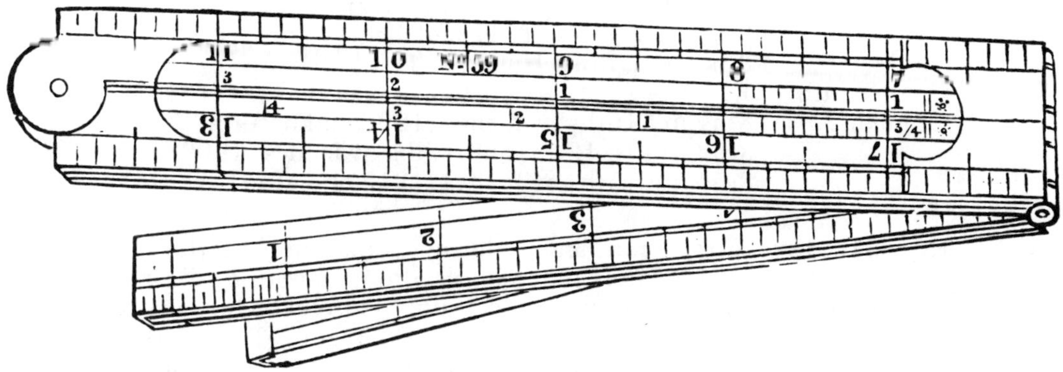

Figure 133: No. 59, 1870 *(Ref. 8)*

60 Carpenters' Rule, 2 Foot, 4 Fold

Offered: 1855 thru 1915
Construction: Double Arch Joint, Full Bound
Material: Boxwood, Brass Hinges & Trim
Width: 1 Inch
Graduations: 8ths and 16ths of Inches (1865 and before)
8ths, 10ths, and 16ths of Inches (1867 thru 1870)
8ths and 16ths of Inches (1872)
8ths, 10ths, and 16ths of Inches (1874 thru 1877)
8ths, 10ths, 12ths, and 16ths of Inches (1879 and after)
Scales: Drafting

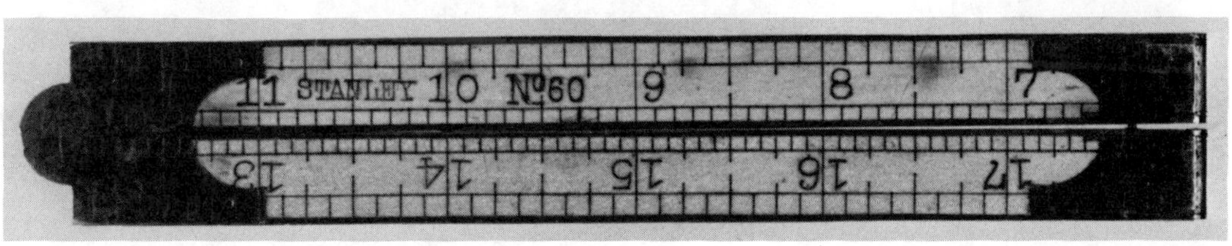

Figure 134: No. 60, 1879–1915 *(Charles and Walter Jacob Collection)*

It is not clear whether the 10ths scale was added to and deleted from the No. 60 rule as often as the catalogues would have us believe, or whether this is simply a misprint in one or more of those catalogues. The latter case would seem the more likely, in view of Stanley's normal rate of changing the product during this period. If so, then the misprint probably occurred in the 1872 catalogue (Ref. 9), and the 10ths scale was a feature of the No. 60 for the entire period from 1867 through 1877.

61 Carpenters' Rule, 2 Foot, 4 Fold

Offered:	1855 thru 1941, 1947 thru 1958
Construction:	Square Joint, Middle Plates Printed Figures (1934 and after) Vertical Figures (1947 and after)
Material:	Boxwood, Brass Hinges & Tips
Width:	1 Inch
Graduations:	8ths and 16ths of Inches (1907 and before) 8ths, 10ths, 12ths, and 16ths of Inches (1909) 8ths and 16ths of Inches (1912 and after)

In 1947 (Ref. 54) there was a conflict in the catalogue between the description of the No. 61 and the illustration; the former describing the rule as having vertical graduations, the latter showing a rule with horizontal ones. This was probably an oversight during preparation of the catalogue, and the description, not the illustration, is almost certainly correct. This was corrected in the 1949 catalogue (Ref. 55), the illustration being changed to correspond to the text.

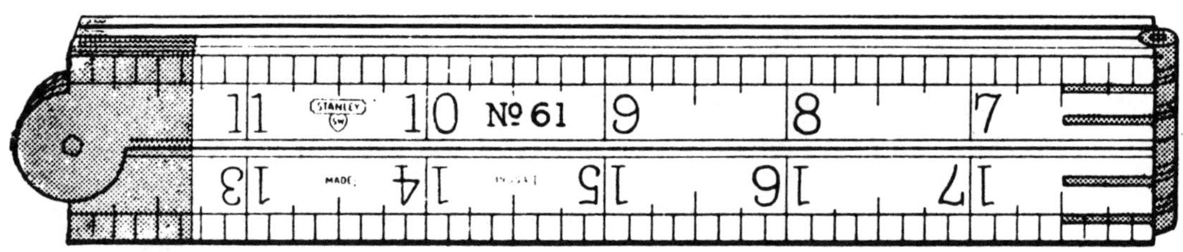

Figure 135a: No. 61, 1929 *(Ref. 41)*

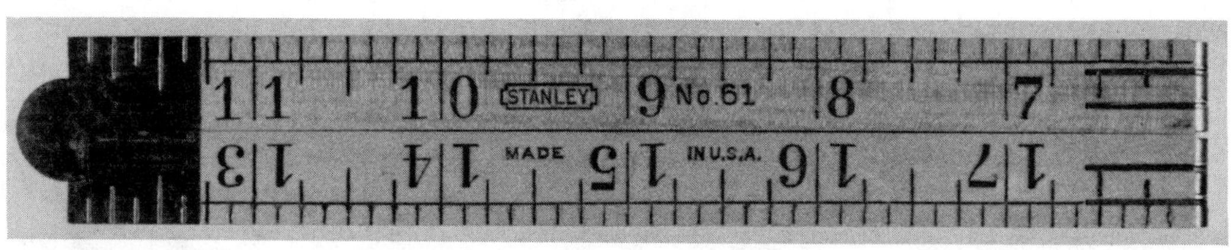

Figure 135b: No. 61, 1934–1941 *(Author's Collection)*

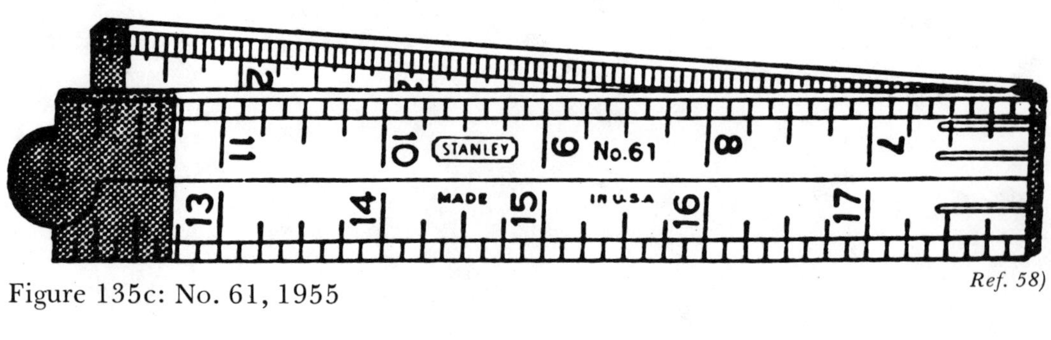

Figure 135c: No. 61, 1955 *Ref. 58)*

In 1942 (Ref. 52) eleven rules were "temporarily" discontinued as part of Stanley's switching to war production, and one of these was the No. 61. In the case of some of the others, this temporary hiatus proved permanent, and they never returned, but this was not so with the No. 61. It was returned to the catalogue in 1947 (Ref. 54), and remained in production for another eleven years.

Heavy Figure Rules

"Heavy Figures" was Stanley's term for the printed graduations and figures which were introduced on several of their more common rules during the 1920's. They were an intermediate stage between the incised graduations and stamped figures then in use, and the extra large and heavy printed figures and graduations of the Nos. 7 and 8 "Blindman's" rules introduced at the turn of the century. Printed markings were standard on the "Zig-Zag" rules which Stanley had been manufacturing since 1900, but except for the Nos. 7 and 8 had not hitherto been used for boxwood rules.

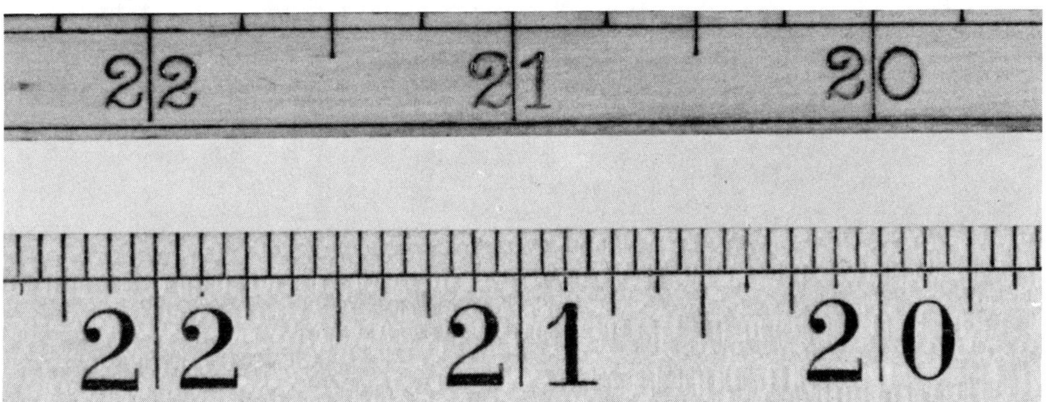

Figure 136: Comparison of Ordinary and "Heavy" Figures *(Author's Collection)*

It is not clear whether this new method of marking was adopted in order to reduce product cost, or because Stanley felt that there was a demand for rules with heavier figures and graduations. It had both advantages and disadvantages. The heavier graduations and figures were certainly easier to read, but the wider graduation lines made it more difficult to transfer a mark from the rule to the work precisely. Additionally, the printed markings, being on the surface of the rule, were more subject to wear (one has only to consider the number of rules with illegible printed markings encountered at flea markets, etc., to be aware of this).

Four rules with "heavy" figures were made, the Nos. 61A, 66¼A, 66½A, and 68A. They were identical to their similarly numbered counterparts, the Nos. 61, 66¼, 66½, and 68, differing only in having printed graduations and figures instead of incised and stamped. The No. 68A was introduced in 1919 (Ref. 33), the No. 66½A in 1923 (Ref. 36), the No. 68A in 1926 (Ref. 39), and the No. 66¼A in 1929 (Ref. 41).

Up until 1932 (Ref. 43) these heavy figure rules were sold in parallel with their incised/stamped counterparts at the same price. Then, in 1934 (Ref. 44), they were "deleted" from the product line. At the same time, however their incised/stamped counterparts (the Nos. 61, etc.) were changed to have heavy printed figures and graduations! In effect, Stanley had actually removed the incised/stamped versions of these rules from the line, and had renumbered the heavy-figure versions to take their place.

Not all boxwood rules were converted to printed markings at this time. The Nos. 18, 53½, 62, etc., continued to be shown in the catalogue as having stamped/incised markings until they were discontinued at various times over the next twenty years.

These four rules were not the only ones to bear printed markings. Most of the eleven rules introduced after 1930 (the Nos. 27,

136, 163, 214T, etc.) were also so marked and graduated, and still more (the Nos. 36½L and 36½R, for example) were switched over during the next two decades.

61A Carpenters' Rule, 2 Foot, 4 Fold

Offered:	1926 thru 1932
Construction:	Square Joint, Middle Plates Extra Heavy Figures
Material:	Boxwood, Brass Hinges & Tips
Width:	1 Inch
Graduations:	8ths and 16ths of Inches

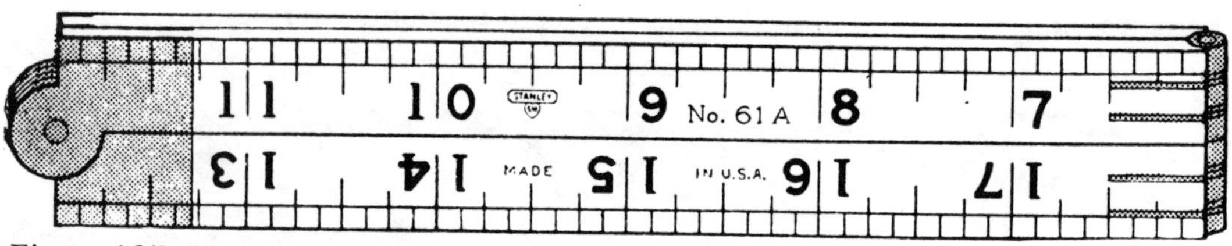

Figure 137: No. 61A, 1929 *(Ref. 41)*

61½ Carpenters' Rule, 2 Foot, 4 Fold

Offered:	1855 thru 1917
Construction:	Square Joint, Middle Plates
Material:	Boxwood, Brass Hinges & Tips
Width:	¾ Inch
Graduations:	8ths and 16ths of Inches

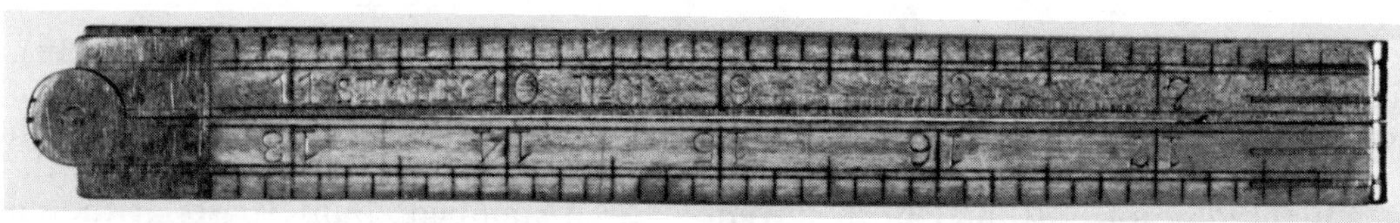

Figure 138: No. 61½, 1879–1917 *(Charles and Walter Jacob Collection)*

62 Carpenters' Rule, 2 Foot, 4 Fold

Offered:	1855 thru 1941
Construction:	Square Joint, Full Bound
Material:	Boxwood, Brass Hinges & Trim
Width:	1 Inch
Graduations:	8ths and 16ths of Inches (1865 and before) 8ths, 10ths, and 16ths of Inches (1867 thru 1877) 8ths, 10ths, 12ths, and 16ths of Inches (1879 and after)
Scales:	Drafting

In 1942 the No. 62, along with many other rules, was discontinued to conserve materials and to make room for more important war production. At that time the price list (Ref. 52) described this cancellation as only "temporary," but this decision was subsequently modified, and when the first postwar catalogue was published in 1947 (Ref. 54) the No. 62 had been dropped permanently.

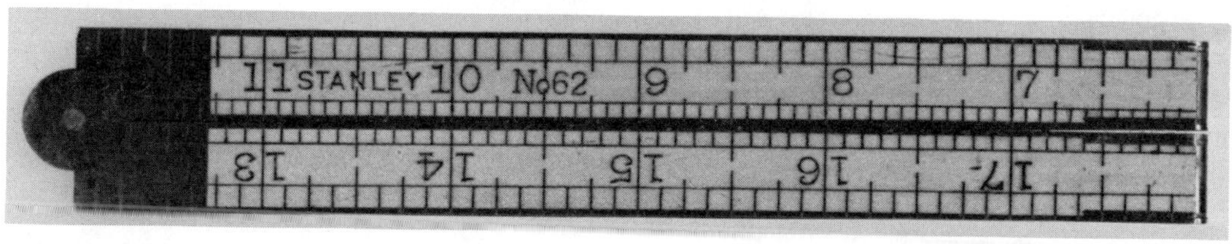

Figure 139: No. 62, 1879–1921 *(Author's Collection)*

62C Carpenters' Caliper Rule, 2 Foot, 4 Fold

Offered: 1910 thru 1932
Construction: Square Joint, Full Bound, Left Hand Caliper
Material: Boxwood, Brass Hinges, Caliper & Trim
Width: 1 Inch
Graduations: Body: 8ths, 10ths, 12ths, and 16ths of Inches
Caliper: 16ths of Inches (1922 and before)
16ths and 32nds of Inches (1925 and after)
Scales: Drafting (1925 and after)

The No. 62C is the most common of the 2 foot, 4 fold caliper rules, and a number of examples have been observed by the author in near new condition. This would tend to indicate that either this rule was continued in production for a number of years after its last appearance in the catalogue in 1932 (Ref. 43), or that when production was suspended a large number of these rules were on hand and they continued to be sold informally until these stocks were expended. Two other facts support these possibilities: the first, that several of these known examples bear the "Made In USA" trademark indicative of post-1932 manufacture, and the second, that the 1942 supplement (Ref. 52) to the 1941 dealers' catalogue listed this rule as one which had been "permanently discontinued" due to the war effort.

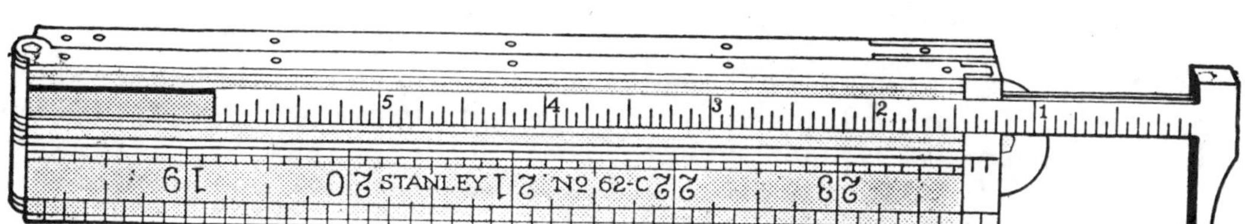

Figure 140a: No. 62C, 1912 *(Ref. 27)*

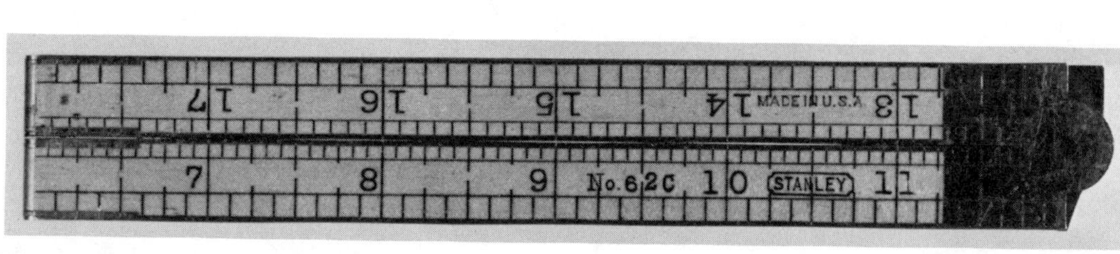

Figure 140b: No. 62C, 1934–1941 *(Author's Collection)*

It is not known whether the No. 62C was still in production in 1941, or if Stanley was simply using up partially finished stocks. In either event, after the end of hostilities, this rule was not one of those which was returned to the product line; like literally dozens of others it had fallen victim to economics and the competition of newer types of rules.

62½ Carpenters' Rule, 2 Foot, 4 Fold
Offered: 1874 thru 1932
Construction: Square Joint, Full Bound
Material: Boxwood, Brass Hinges & Trim
Width: ¾ Inch
Graduations: 8ths, 10ths, 12ths, and 16ths of Inches

The 1929 dealers' catalogue (Ref. 41) describes this rule as graduated in 8ths, 10ths, and 16ths of inches, with no mention of 12ths. This is probably a misprint; all other catalogue references list all four scales, and all known examples of this not uncommon rule exhibit them.

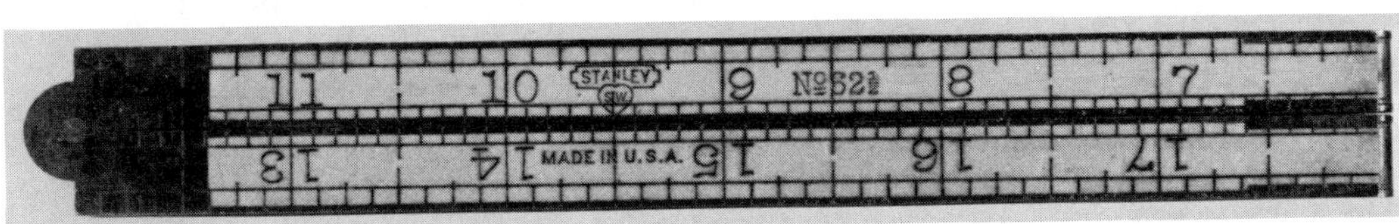

Figure 141: No. 62½, 1922–1932 *(Author's Collection)*

63 Carpenters' Rule, 2 Foot, 4 Fold
Offered: 1855 thru 1941, 1947 thru 1950
Construction: Square Joint, Edge Plates
10ths Scale on Edge (1898 and before)
10ths Scale on Surface (1909 and after)
Material: Boxwood, Brass Hinges & Tips
Width: 1 Inch
Graduations: 8ths, 10ths, and 16ths of Inches (1877 and before)
8ths, 10ths, 12ths, and 16ths of Inches (1879 and after)
Scales: Drafting

In 1942 the No. 63, along with many other rules, was discontinued to conserve materials and to make room for more important war production. The price list (Ref. 52) described this cancellation as only "temporary," and although some other rules so canceled were never returned to production, this was not the case with the No. 63, and it was back in the product line when the first postwar catalogue was published in 1947 (Ref. 54).

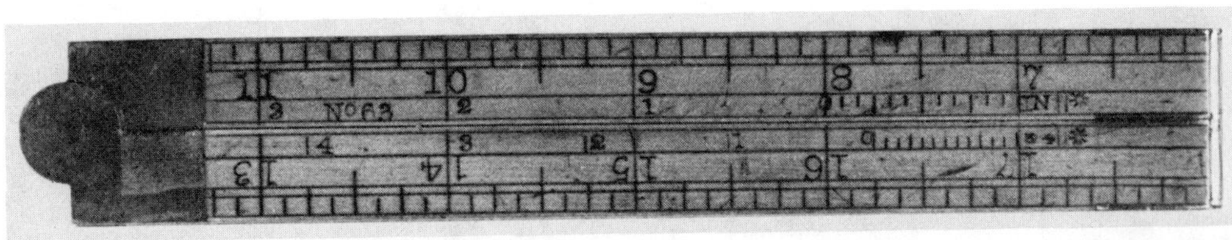

Figure 142: No. 63, 1854–1857 *(Paul Kebabian Collection)*

63½ Carpenters' Rule, 2 Foot, 4 Fold

Offered:	1860 thru 1932
Construction:	Square Joint, Edge Plates 10ths Scale on Edge (1860)
Material:	Boxwood, Brass Hinges & Tips
Width:	¾ Inch
Graduations:	8ths, 10ths, and 16ths of Inches

The 1867 through 1872 catalogues (Refs. 7 thru 9) describe the No. 63½ as being of "Extra Quality." It is not clear what is meant by this phrase, and since there is no feature by which No. 63½ rules of this period can be distinguished, examination of samples has not been able to tell us.

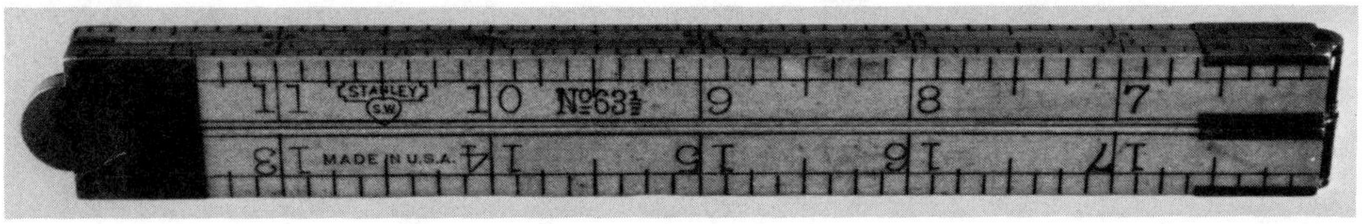

Figure 143: No. 63½, 1922–1932 *(Author's Collection)*

64 Carpenters' Rule, 1 Foot, 4 Fold

Offered:	1855 thru 1932
Construction:	Square Joint, Edge Plates
Material:	Boxwood, Brass Hinges & Tips
Width:	½ or ⅝ Inch (1860 and before) ⅝ Inch (1862 and after)
Graduations:	8ths and 16ths of Inches

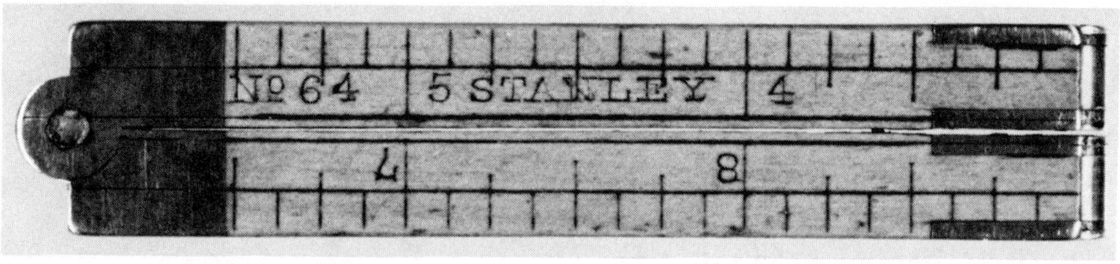

Figure 144: No. 64, 1879–1921 *(Eugene Frankio Collection)*

65 Carpenters' Rule, 1 Foot, 4 Fold

Offered:	1855 thru 1932
Construction:	Square Joint, Middle Plates
Material:	Boxwood, Brass Hinges & Tips
Width:	1/2 or 5/8 Inch (1860 and before)
	5/8 Inch (1862 and after)
Graduations:	8ths and 16ths of Inches

Figure 145: No. 65, 1854–1921 *(Author's Collection)*

65½ Carpenters' Rule, 1 Foot, 4 Fold

Offered:	1855 thru 1932
Construction:	Square Joint, Full Bound
Material:	Boxwood, Brass Hinges & Trim
Width:	1/2 or 5/8 Inch (1860 and before)
	5/8 Inch (1862 and after)
Graduations:	8ths an 16ths of Inches

Figure 146: No. 65½, 1854–1921 *(Roger K. Smith Collection)*

66 Yard Measure, 3 Foot, 4 Fold

Offered:	1855 thru 1917
Construction:	Square Joint (1857 and before)
	Arch Joint (1859 and after)
	Middle Plates
Material:	Boxwood, Brass Hinges & Tips
Width:	1 Inch
Graduations:	16ths of Inches (Outside)
	Fractions of a Yard (Inside)

Although never mentioned in any catalogue, the No. 66 originally had a square main joint. The 1855 price list (Ref. 1) makes no mention of the type of joint, and all subsequent catalogues describe it as an arch joint, but at least one example (see Fig. 147b) has been observed with a square joint. This rule bears an A. Stanley & Co. trademark, indicative of pre-1858 manufacture, hence the changeover from square to arch joint must have therefore occurred prior to that time.

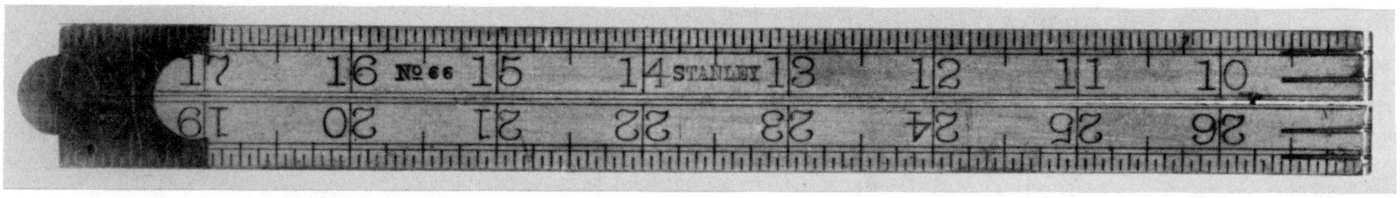

Figure 147a: No. 66, 1879–1911 *(Author's Collection)*

The No. 66 is unique among Stanley yard sticks in that it has two sets of yard graduations. One set runs from right to left on the upper edge of the rule, and the other from left to right on the lower edge. Both are graduated in 8ths of a yard, with additional marks at the $^1/_{32}$ and $^1/_{16}$ yard points.

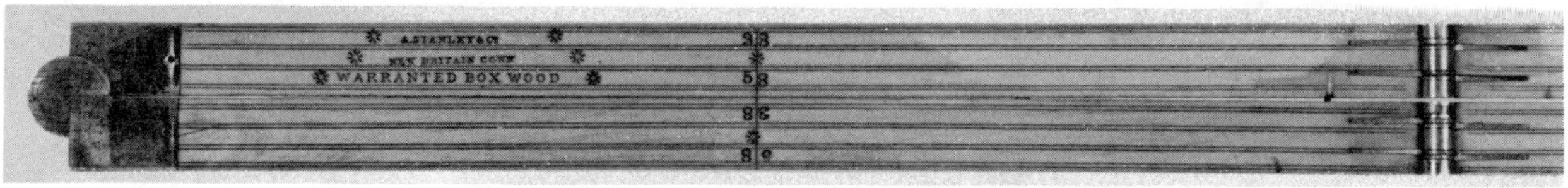

Figure 147b: Yard Marking on the Inside of a No. 66, 1854–1857 *(Bud Steere Collection)*

66¼ Carpenters' Rule, 3 Foot, 4 Fold

Offered:	1910 thru 1941
Construction:	Arch Joint, Edge Plates Printed Figures (1934 and after)
Material:	Boxwood, Brass Hinges & Tips
Width:	1 Inch
Graduations:	8ths and 16ths of Inches

In 1942, the No. 66¼, along with many other rules, was discontinued to conserve materials and to make room for more important war production. At that time the price list (Ref. 52) described this cancellation as only "temporary," but this decision was subsequently modified, and when the first postwar catalogue was published in 1947 (Ref. 54) the No. 66¼ had been dropped permanently.

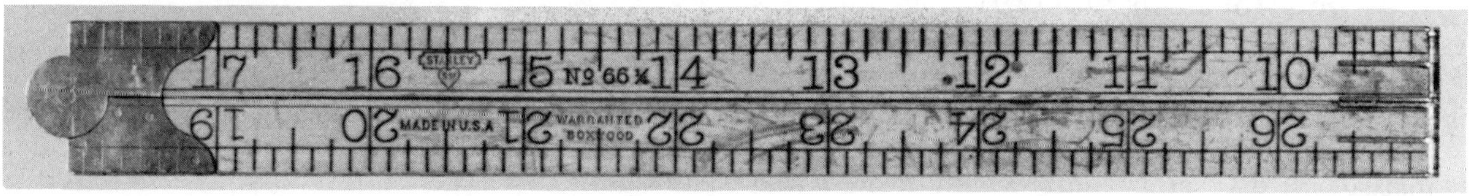

Figure 148: No. 66¼, 1922–1932 *(Author's Collection)*

66¼A Carpenters' Rule, 3 Foot, 4 Fold

Offered: 1929
Construction: Arch Joint, Edge Plates
Extra Heavy (Printed) Figures
Material: Boxwood, Brass Hinges & Tips
Width: 1 Inch
Graduations: 8ths and 16ths of Inches

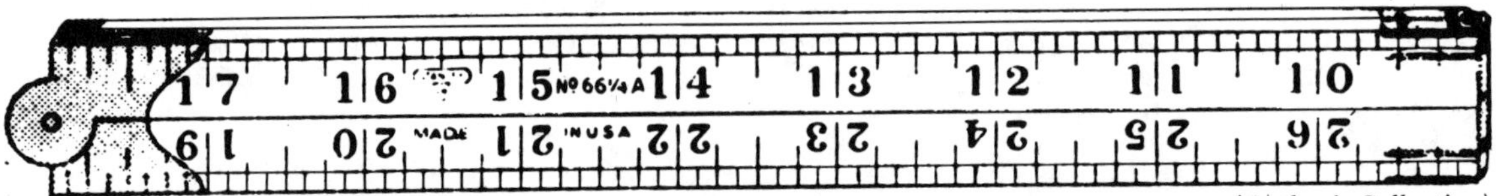

Figure 149: No. 66¼A, 1929 *(Author's Collection)*

66½ Carpenters' Rule, 3 Foot, 4 Fold

Offered: 1870 thru 1957
Construction: Arch Joint, Middle Plates (1941 and before)
Square Joint, Middle Plates (1947 and after)
Printed Figures (1934 and after)
Material: Boxwood, Brass Hinges & Tips
Width: 1 Inch
Graduations: 8ths and 16ths of Inches

The No. 66½ was the only one of the three remaining 3 foot, 4 fold rules to survive World War II (the No. 66¼ had been canceled in 1941 [Ref. 51], and the No. 66¾ the next year [Ref. 52]).

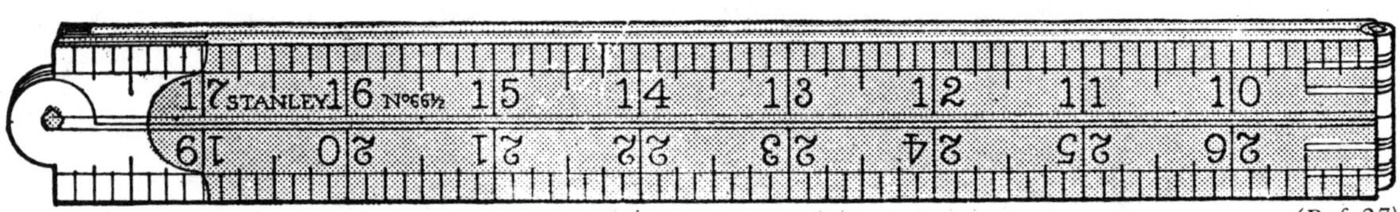

Figure 150a: No. 66½, 1912 *(Ref. 27)*

Sometime between 1941 and 1947, this rule was redesigned to have a square joint, instead of the arch joint which it had had hitherto. This change was presumably made to reduce production labor during the war, and was retained afterward to keep costs down.

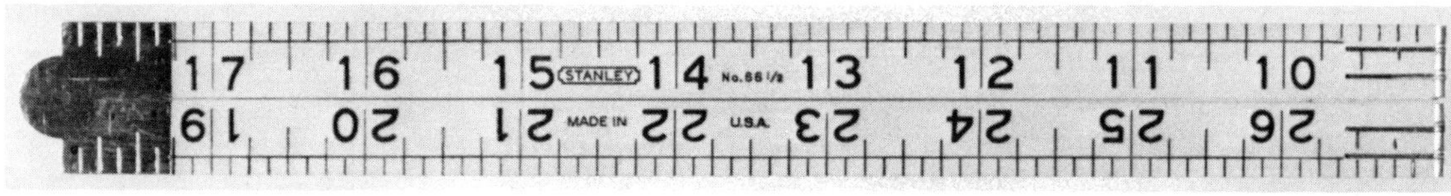

Figure 150b: No. 66½, 1947–1957 *(Charles and Walter Jacob Collection)*

66½A Carpenters' Rule, 3 Foot, 4 Fold

Offered:	1923 thru 1932
Construction:	Arch Joint, Middle Plates Extra Heavy (Printed) Figures
Material:	Boxwood, Brass Hinges & Tips
Width:	1 Inch
Graduations:	8ths and 16ths of Inches

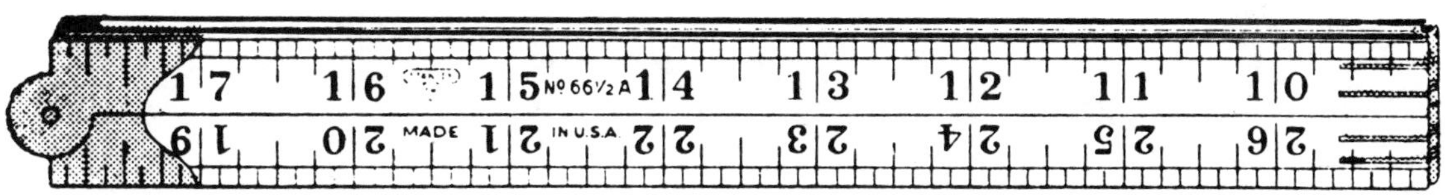

Figure 151: No. 66½A, 1929 *(Ref. 41)*

66¾ Carpenters' Rule, 3 Foot, 4 Fold

Offered:	1902 thru 1941
Construction:	Arch Joint, Full Bound Printed Figures (1936 and after)
Material:	Boxwood, Brass Hinges & Trim
Width:	1 Inch
Graduations:	8ths and 16ths of Inches

In 1942, the No. 66¾, along with many other rules, was discontinued to conserve materials and to make room for more important war production. At that time the price list (Ref. 52) described this cancellation as only "temporary." but this decision was subsequently modified, and when the first post-war catalogue was published in 1947 (Ref. 54) the No. 66¾ had been dropped permanently.

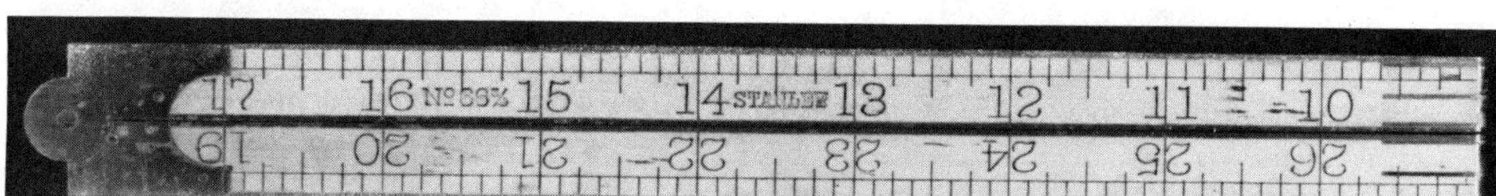

Figure 152: No. 66¾, 1902–1921 *(James Hill Collection)*

67 Carpenters' Rule, 2 Foot, 4 Fold
Offered: 1855 thru 1917
Construction: Round Joint, Middle Plates
Material: Boxwood, Brass Hinges & Tips
Width: 1³/₈ Inch
Graduations: 8ths and 16ths of Inches

Figure 153: No. 67, 1879–1917 *(William Baader Collection)*

68 Carpenters' Rule, 2 Foot, 4 Fold
Offered: 1855 thru 1958
Construction: Round Joint, Middle Plates
Printed Figures (1934 and after)
Vertical Figures (1947 and after)
Material: Boxwood, Brass Hinges & Tips
Width: 1 Inch
Graduations: 8ths and 16ths of Inches

The No. 68 was the basic rule in the Stanley line of rules for over 100 years. Having a round joint and middle plates, and being graduated only in 8ths and 16ths, it was the least elaborate and least expensive rule they made (and probably the best selling). The design of this rule stayed remarkably stable over its entire lifetime, only the method of marking the graduations being changed in 1934 (Ref. 44), and the figures changing from horizontal to vertical in 1947 (Ref. 54).

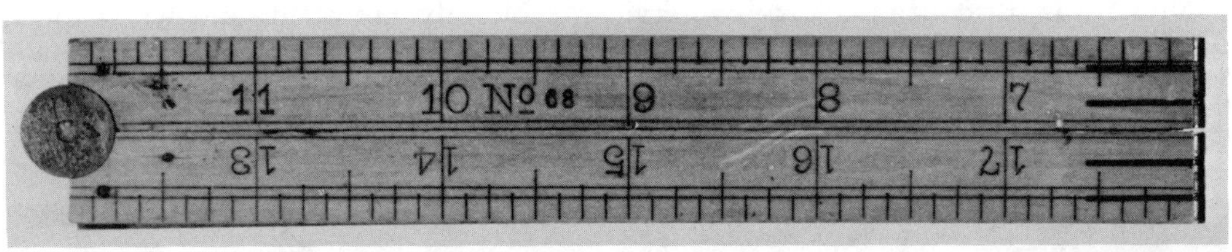

Figure 154a: No. 68, 1858–1898 *(Author's Collection)*

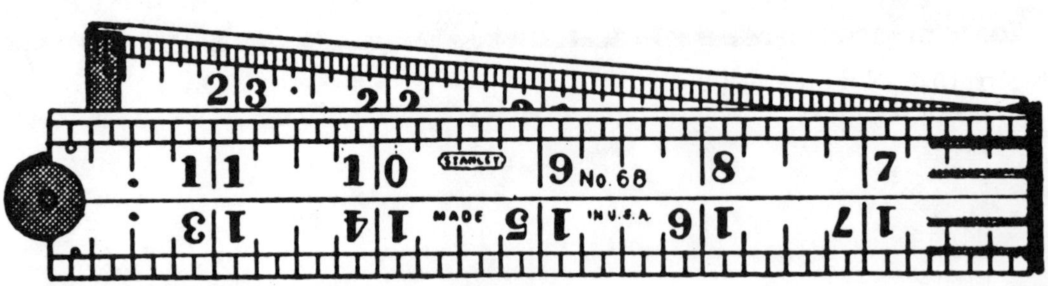

Figure 154b: No. 68, 1941 *(Ref. 51)*

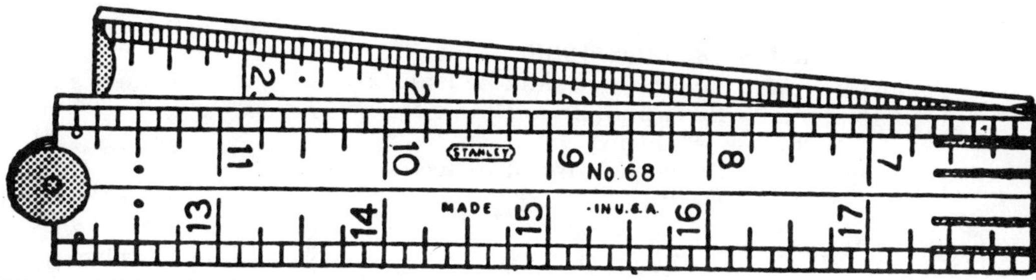

Figure 154c: No. 68, 1957 *(Ref. 59)*

68A Carpenters' Rule, 2 Foot, 4 Fold

Offered:	1919 thru 1932
Construction:	Round Joint, Middle Plates Extra Heavy Figures
Material:	Boxwood, Brass Hinges & Tips
Width:	1 Inch
Graduations:	8ths and 16ths of Inches

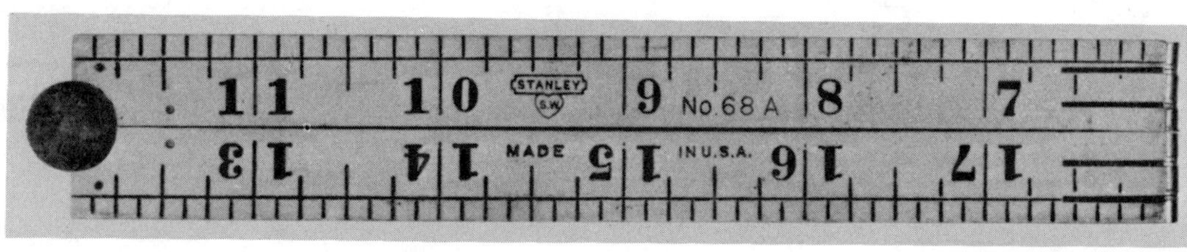

Figure 155: No. 68A, 1922–1932 *(Author's Collection)*

69 Carpenters' Rule, 1 Foot, 4 Fold

Offered:	1855 thru 1932
Construction:	Round Joint, Middle Plates
Material:	Boxwood, Brass Hinges & Tips
Width:	½ or ⅝ Inch (1860 and before) ⅝ Inch (1862 and after)
Graduations:	8ths and 16ths of Inches

Figure 156: No. 69, 1854–1921 *(Author's Collection)*

70 Carpenters' Rule, 2 Foot, 4 Fold

Offered: 1855 thru 1938
Construction: Square Joint, Middle Plates
Material: Boxwood, Brass Hinges & Tips
Width: 1$^3/_8$ Inch
Graduations: 8ths and 16ths of Inches
Scales: Drafting

In 1938 (Ref. 47) the No. 70, with incised graduations and stamped figures, was discontinued, replaced by the recently introduced No. 70E. The No. 70E was basically the same rule as the No. 70, but had printed figures and graduations, and was graduated left to right ("English" graduations). Other rules were also being switched from incised to printed graduations at about this same time, but the No. 70/70E was the only one which was renumbered in the process; presumably it was done in this case because of the change to left to right graduations.

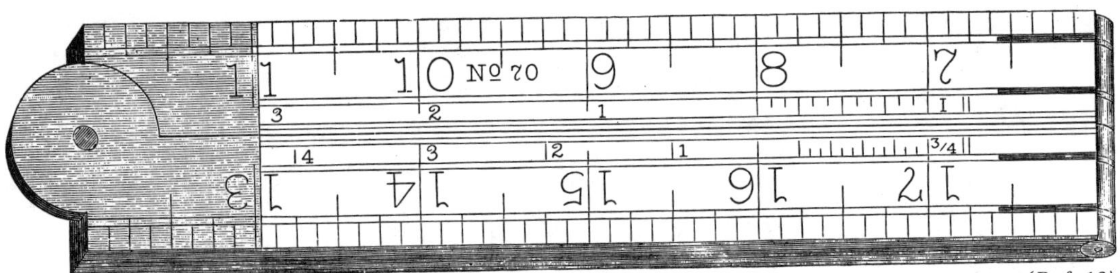

(Ref. 13)

Figure 157: No. 70, 1884

70E Carpenters' ("Blindman's") Rule, 2 Foot, 4 Fold

Offered: 1937 thru 1943
Construction: Square Joint, Middle Plates
Extra Large & Heavy Figures, and Extra Heavy Graduations
Left-to-Right ("English") Markings
Material: Boxwood, Brass Hinges & Tips
Width: 1$^3/_8$ Inch
Graduations: 8ths, 10ths, 12ths, and 16ths of Inches
Scales: Drafting

In 1937 (Ref. 46), in an effort to reduce manufacturing cost, a less expensive version of the No. 70, the No. 70E, was added to the line. This new rule was identical in construction to the No. 70, but replaced the incised graduations and figures with printed ones. A second difference, unrelated to cost, was that the figures on the new rule were arranged left-to-right ("English" graduations) instead of the usual right-to-left arrangement. During the mid-1920's Stanley had begun to mark more rules with left to right markings (the yard sticks were switched over at about this time), and the new No. 70E was also one of these.

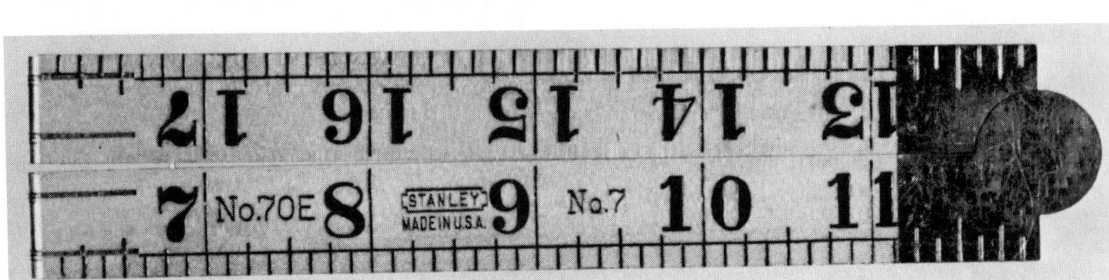

(Charles and Walter Jacob Collection)

Figure 158: No. 70E, 1937–1943

A number of rules structurally identical to the No. 70E, but marked with Blindman's markings, have been observed which are marked both No. 70E and No. 7. These must date from the period 1941–1943, after the No. 7 had drafting scales added to its inside surface. With this change, the only remaining difference between the No. 7 and the No. 70E was their style of marking (the No. 70E had heavy, instead of Blindman's figures). Apparently when this change was made in the No. 7 it was decided to merge the two rules, selling the No. 7 as a No. 70E as well. Two years later, the No. 70E was discontinued as a separate identity for this rule, and it resumed its previous single identity as the No. 7.

71 Carpenters' Rule, 2 Foot, 4 Fold

Offered:	1855
Construction:	Square Joint, Middle Plates, Bitted
Material:	Boxwood, Brass Hinges & Trim
Width:	Not Known, but probably 1-3/8 Inch
Graduations:	Not Known

71 Carpenters' Rule, 3 thru 6 Foot

Offered:	1870 thru 1943, 1949 thru 1957
Construction:	Flat
Material:	Maple, Brass Tips
Length:	4 Foot (1917 and before) 3, 4, 5, or 6 Foot (1919 thru 1942) 6 Foot (1943) 4 Foot (1949 and after)
Width:	1¼ Inch (1909 and before, 1927 thru 1943) 1½ Inches (1919 thru 1926, 1949 and after)
Graduations:	8ths of Inches 10ths of a Foot (1917 and before) 16ths of Inches (1919 and after)

This rule was called a Wood Measure until 1922 (Ref. 34). It is not known if there is any significance to that particular name, or whether this was just a minor nomenclatural variation.

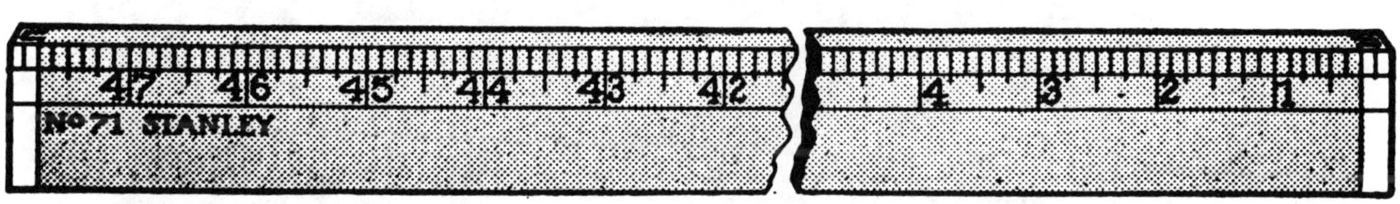

Figure 159a: No. 71, 4 Foot, 1919 *(Ref. 33)*

The No. 71 was not listed in the 1872 catalogue (Ref. 9). This may be indicative of some uncertainty as to whether to continue marketing the rule or not, or may have been an accidental omission.

During its lifetime, the No. 71 was offered in several different arrangements of graduations. Prior to 1926, the graduations were on the upper edge of the rule, from right to left; from 1926 through 1941, on the lower edge, from left to right; after World War II, both types were provided. Figs. 159a thru 159c show the three arrangements.

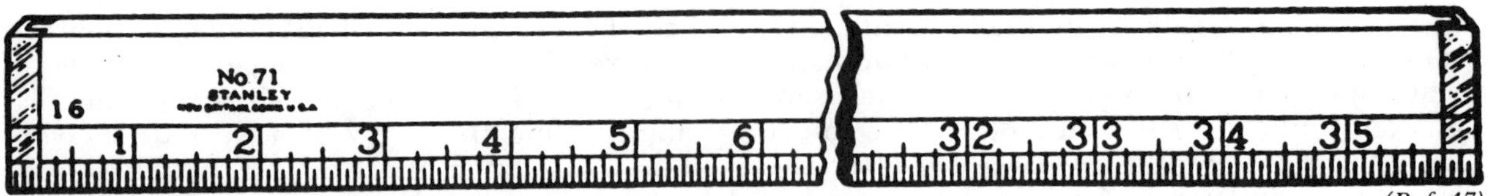

Figure 159b: No. 71, 3 Foot, 1938 *(Ref. 47)*

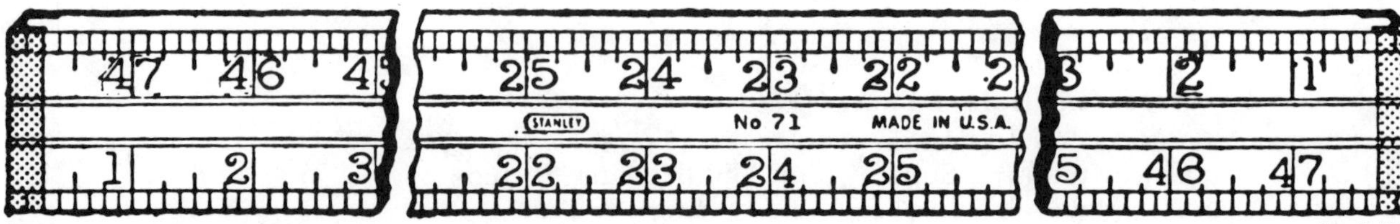

Figure 159c: No. 71, 4 Foot, 1949 *(Ref. 55)*

The No. 71 ceased to be offered sometime between 1943 (Ref. 53) and 1947 (Ref. 54), probably during 1944–1945 in response to war production necessities. Production was not resumed until 1949 (Ref. 55), which was unusual; all other rules which returned to production after the end of the war did so in 1947. Why the No. 71 was not offered again until two years later is not known.

72 Carpenters' Rule, 2 Foot, 4 Fold

Offered:	1855 thru 1932
Construction:	Square Joint, Edge Plates 10ths Scale on Edge (1860 and before) 10ths Scale on Surface (1919 and after)
Material:	Boxwood, Brass Hinges & Tips
Width:	1$^{3}/_{8}$ Inch
Graduations:	8ths, 10ths, and 16ths of Inches
Scales:	Drafting

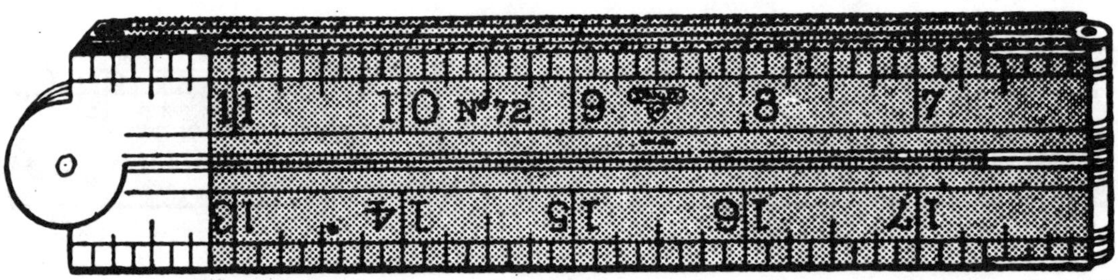

Figure 160: No. 72, 1932 *(Ref. 43)*

72½ Carpenters' Rule, 2 Foot, 4 Fold

Offered:	1855 thru 1932
Construction:	Square Joint, Full Bound
Material:	Boxwood, Brass Hinges & Trim
Width:	1$^{3}/_{8}$ Inch
Graduations:	8ths, 10ths, and 16ths of Inches
Scales:	Drafting

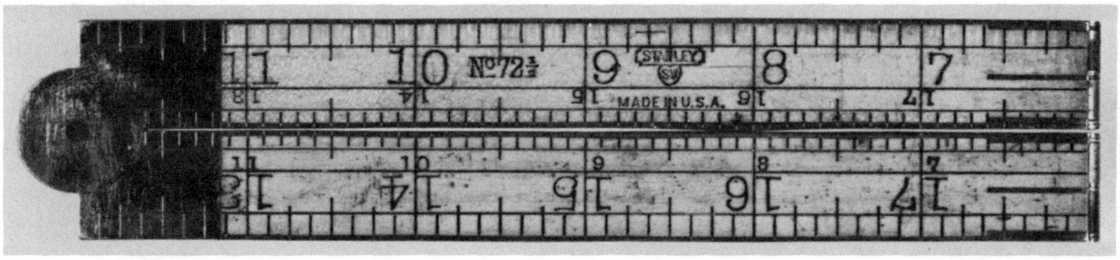

Figure 161: No. 72½, 1922–1932 *(Author's Collection)*

73 Carpenters' Rule, 2 Foot, 4 Fold

Offered:	1855 thru 1932
Construction:	Arch Joint, Middle Plates
Material:	Boxwood, Brass Hinges & Tips
Width:	1⅜ Inch
Graduations:	8ths and 16ths of Inches (1872 and before) 8ths, 10ths, and 16ths of Inches (1874 and after)
Scales:	Drafting

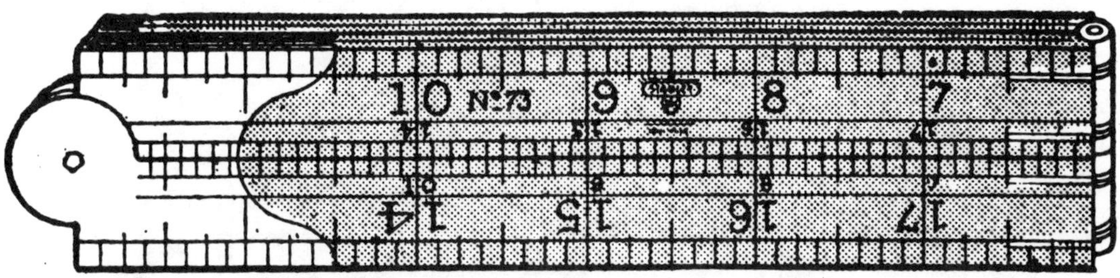

Figure 162: No. 73, 1927 *(Ref. 40)*

73¼ Carpenters' Rule, 2 Foot, 4 Fold

Offered:	1929, 1937
Construction:	Arch Joint, Middle Plates Left-to-Right ("English") Markings
Material:	Boxwood, Brass Hinges & Tips
Width:	1½ Inch
Graduations:	8ths, 10ths, 12ths, and 16ths of Inches

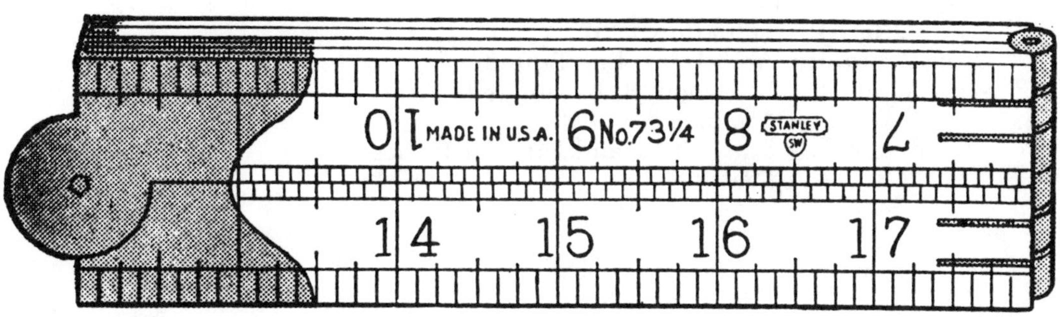

Figure 163: No. 73¼, 1929 *(Ref. 41)*

The No. 73¼ was another of the rules with left to right graduations introduced by Stanley during the 1920's. It was introduced in 1929 (Ref. 41), along with two other similar rules (the Nos. 173E and 173¼E). All three were removed from the catalogue the following year (Ref. 42), probably due to the onset of the Depression of 1929–1939.

A second attempt was made to introduce this rule in 1937 (Ref. 46), but must have been equally unsuccessful; in 1938 (Ref. 47) it had been discontinued once more, this time never to reappear.

74 Carpenters' Rule, 2 Foot, 4 Fold
Offered: 1855
Construction: Arch Joint, Middle Plates, Bitted
Material: Boxwood, Brass Hinges & Trim
Width: "Broad"; probably 1³/₈ Inch
Graduations: Not Known

75 Carpenters' Rule, 2 Foot, 4 Fold
Offered: 1855 thru 1932
Construction: Arch Joint, Edge Plates
10ths Scale on Edge (1927 and before)
10ths Scale on Surface (1929 and after)
Material: Boxwood, Brass Hinges & Tips
Width: 1³/₈ Inch
Graduations: 8ths, 10ths, and 16ths of Inches
Scales: Drafting

Figure 164: No. 75, 1879–1911 *(Author's Collection)*

76 Carpenters' Rule, 2 Foot, 4 Fold
Offered: 1855 thru 1932
Construction: Arch Joint, Full Bound
Material: Boxwood, Brass Hinges & Trim
Width: 1³/₈ Inch
Graduations: 8ths, 10ths, and 16ths of Inches
Scales: Drafting

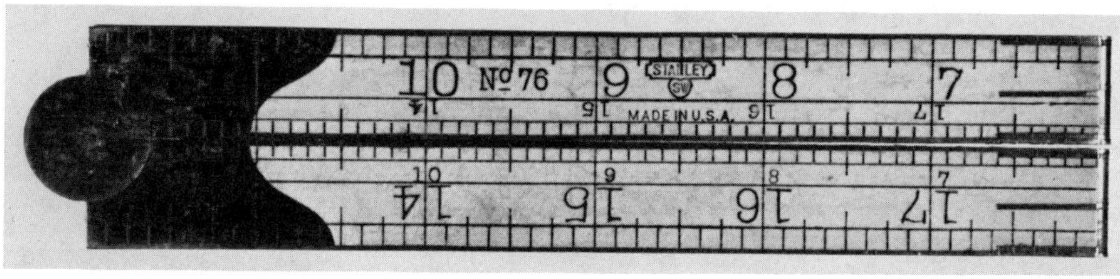

Figure 165: No. 76, 1922–1932 *(Roger K. Smith Collection)*

76C Carpenters' Caliper Rule, 2 Foot, 4 Fold

Offered: 1910 thru 1917
Construction: Arch Joint, Full Bound, Left Hand Caliper
Material: Boxwood, Brass Hinges, Caliper & Trim
Width: $1^3/_8$ Inch
Graduations: Body: 8ths, 10ths, and 16ths of Inches
Caliper: 32nds of Inches

The No. 76C was identical to the No. 76, but had a 6 inch left hand caliper slide in one of the tip sticks. See the illustration of the No. 83C for the general configuration of this rule.

77 Carpenters' Rule, 2 Foot, 4 Fold

Offered: 1855 thru 1915
Construction: Double Arch Joint
Edge Plates (1855)
Middle Plates, Bitted (1865 thru 1909)
Edge Plates (1912 and after)
Material: Boxwood, Brass Hinges & Trim
Width: $1^3/_8$ Inch
Graduations: 8ths, 10ths, and 16ths of Inches
Scales: Drafting

The style of plates in the middle joints of the No. 77 was changed twice during the sixty years that this rule was produced. Originally made with edge plates, it was later changed to have middle plates with bits, and then still later back to edge plates again.

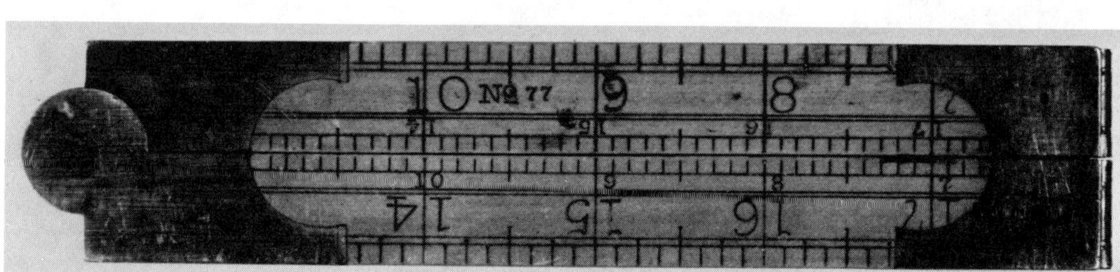

Figure 166: No. 77, 1865–1909 *(Author's Collection)*

The 1855 A. Stanley & Co. Price List (Ref. 1) describes the No. 77 as having "edge Plates."

From 1859 (Ref. 2) to 1862 (Ref. 5) the catalogues made no reference at all to the plates of the middle joints. In these years this would ordinarily have meant that the rule had middle plates and no bits.

Beginning in 1865 (Ref. 6) the catalogues began describing the rule as being "bitted." This almost certainly implies that the rule now had middle plates; a rule with edge plates would have no need of separate bitting. This description prevailed until 1909 (Ref. 25).

It is not known why this change was made, but some speculation is possible. In 1855 the new A. Stanley & Co. was in the process of consolidating the different lines of rules previously manufactured separately by Thomas Conklin and Seth Savage. It may be that one or the other had a stock of broad, double arch joint, edge plated rules on hand at the time of the merger, and they were accordingly listed in the 1855 price list, but that at the same time it was being decided to make this class of rule with middle plates and bits, and the description changed as soon as the old stock ran out (1865).

In 1912 (Ref. 27) through 1915 (Ref. 31), the final three years of manufacture of the No. 77, the rule was changed back to have edge plates again. In 1909 (Ref. 25) only two bitted 4 fold rules were still in production, the Nos. 59 and 77. By 1912 (Ref. 27) the No. 59 had been discontinued. It was apparantly decided at the same time to switch the No. 77 back to edge plates, in order to avoid the expense of having to fit bits to only a single, low-volume product.

78 Carpenters' Rule, 2 Foot, 4 Fold

Offered:	1855 thru 1915
Construction:	Double Arch Joint, Half Bound
Material:	Boxwood, Brass Hinges & Trim
Width:	$1^3/_8$ Inch
Graduations:	8ths, 10ths, and 16ths of Inches
Scales:	Drafting

The No. 78 was not listed in the 1860 price list (Ref. 3), but this was probably an error in the list, and not an attempt to discontinue the rule. The price list was completely reset in that year, and again in 1862 (Ref. 5), when the No. 78 was reincluded, and it is reasonable to assume that this was a printer's error which was not detected until after the list was in circulation.

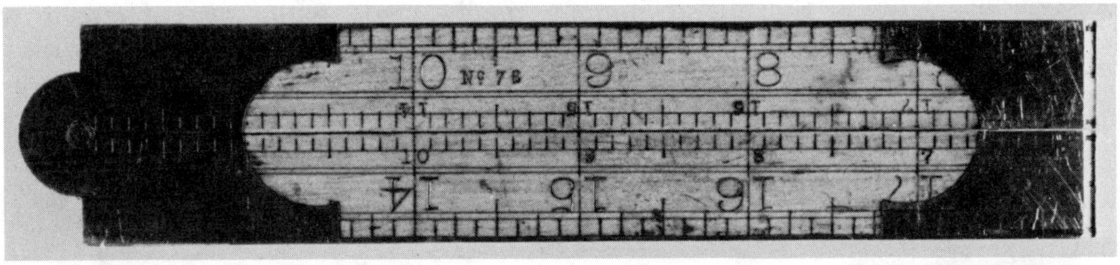

Figure 167: No. 78, 1858–1898 *(Roger K. Smith Collection)*

The production history of the No. 78 would tend to support this supposition about a printing error. The rule was apparently reasonably popular; it was in production for more than fifty years after this time, and was only dropped after 1915 (Ref. 31) as a result of the production demands created by World War I.

78½ Carpenters' Rule, 2 Foot, 4 Fold

Offered:	1855 thru 1932
Construction:	Double Arch Joint, Full Bound
Material:	Boxwood, Brass Hinges & Trim
Width:	1 3/8 Inch
Graduations:	8ths, 10ths, and 16ths of Inches
Scales:	Drafting

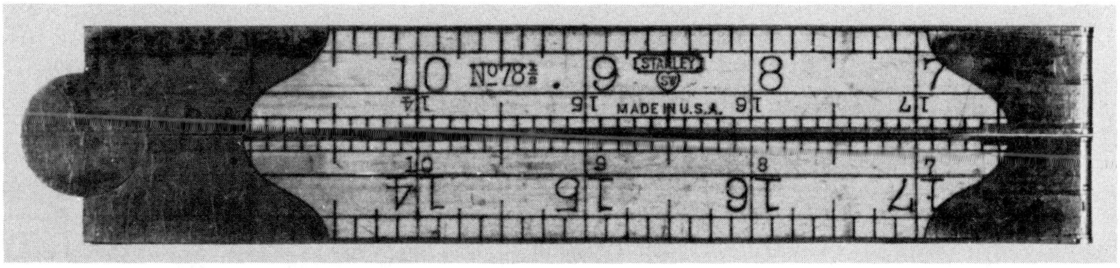

Figure 168: No. 78½, 1922–1932 *(Roger K. Smith Collection)*

79 Carpenters' Rule, 2 Foot, 4 Fold

Offered:	1855 thru 1915
Construction:	Square Joint, Edge Plates
Material:	Boxwood, Brass Hinges & Tips
Width:	1 3/8 Inch
Graduations:	16ths of Inches (1872 and before) 12ths and 16ths of Inches (1874 and after)
Scales:	Drafting
Other Markings:	Board Measure Tables

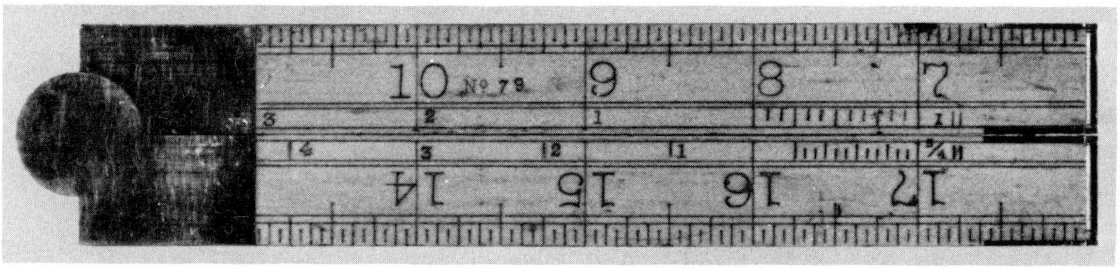

Figure 169: No. 79, 1874–1898 *(Author's Collection)*

80 Carpenters' Rule, 2 Foot, 4 Fold

Offered:	1855 thru 1859
Construction:	Square Joint, Full Bound
Material:	Boxwood, Brass Hinges & Trim
Width:	"Broad," probably 1 3/8 Inch
Graduations:	Not Known
Other Markings:	Board Measure Tables

80 Saddlers' Rule, 3 Foot

Offered:	1884 thru 1943
Construction:	"Capped Ends" (1917 and before)
Material:	Maple, Brass Tips
Width:	1½ Inch
Graduations:	8ths and 16ths of Inches

The No. 80 was the last of the four rules which were retained when Stanley discontinued the rest of the Stearns line of wood rules in 1884 (Ref. 13) (the others were the Nos. 13, 30, and 31). Stanley had not offered a saddlers' rule prior to that time, and the catalogue description of this "new" extra wide 3 foot rule corresponded exactly with that of the Stearns No. 80, which was listed for the last time in 1879 (Ref. 12). It is highly probable that this was not a coincidence, but was simply the transfer to the Stanley product line of a rule too popular to be eliminated along with the Stearns name.

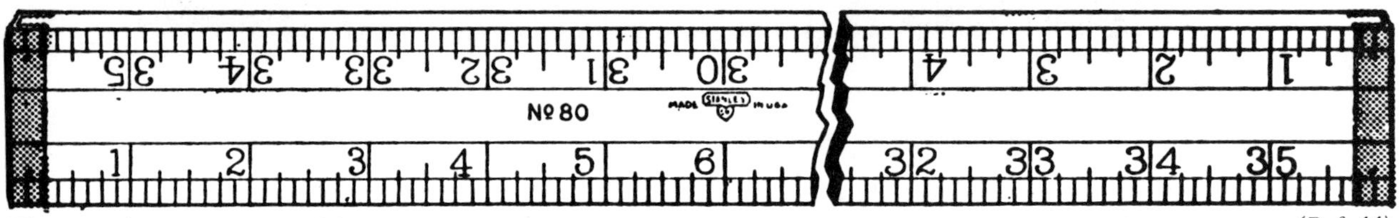

Figure 170: No. 80, 1934 *(Ref. 44)*

This type of rule was known as a "Saddlers'" rule because it was of a length and width suitable for use by the leather workers who made harness and saddles. A 2 foot rule was not long enough for their purposes; many of the pieces which go into harness or saddles are larger than this in at least one dimension. An extra wide rule is also desirable for this work for a different reason. Saddlers would frequently use a rule to guide the knife while cutting leather; leather is tough, and the high forces required to cut it would tend to deflect any 3 foot rule which was only ¾ or 1 inch wide.

81 Carpenters' Rule, 2 Foot, 4 Fold

Offered:	1855 thru 1915
Construction:	Arch Joint, Edge Plates
Material:	Boxwood, Brass Hinges & Tips
Width:	1⅜ Inch
Graduations:	16ths of Inches (1872 and before)
	12ths and 16ths of Inches (1874 and after)
Scales:	Drafting
Other Markings:	Board Measure Tables

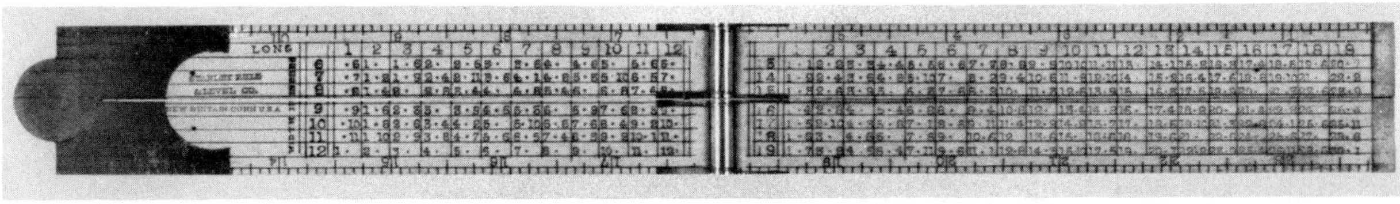

Figure 171: No. 81, 1874–1898 *(Author's Collection)*

82 Carpenters' Rule, 2 Foot, 4 Fold

Offered:	1855 thru 1917
Construction:	Arch Joint, Full Bound
Material:	Boxwood, Brass Hinges & Trim
Width:	1 3/8 Inch
Graduations:	16ths of Inches (1872 and before)
	12ths and 16ths of Inches (1874 and after)
Scales:	Drafting
Other Markings:	Board Measure Tables

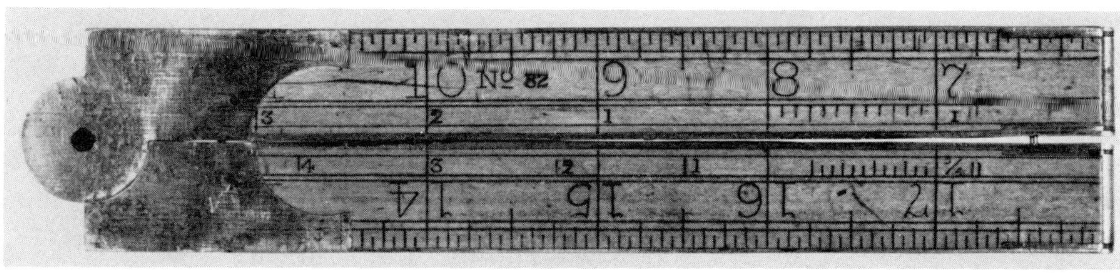

Figure 172: No. 82, 1874—1898 *(Author's Collection)*

83 Carpenters' Rule With Slide, 2 Foot, 4 Fold

Offered:	1859 thru 1915	
Construction:	Arch Joint, Edge Plates, Slide in One Leg	
Material:	Boxwood, Brass Hinges, Slide, & Tips	
Width:	1 3/8 Inch	
Graduations:	Body:	8ths, 10ths, 12ths, and 16ths of Inches (1874 and before)
		8ths, 12ths, and 16ths of Inches (1877 and after)
		100ths of a Foot (On Outside Edge)
	Slide:	8ths of Inches (1874 and after)
Scales:	Body:	Gunter's (1872 and before)
		Octagonal
	Slide:	Gunter's (1872 and before)

The No. 83 is unique among Stanley rules, in that it is the only 4 fold rule ever provided with a Gunter's slide. Such rules were uncommon even among English makers (Refs. 16 and 19), and were very rare in the United States. The slide provided was similar to that of a caliper, but without the head, and had small thumbnail depressions at each end to aid in positioning it.

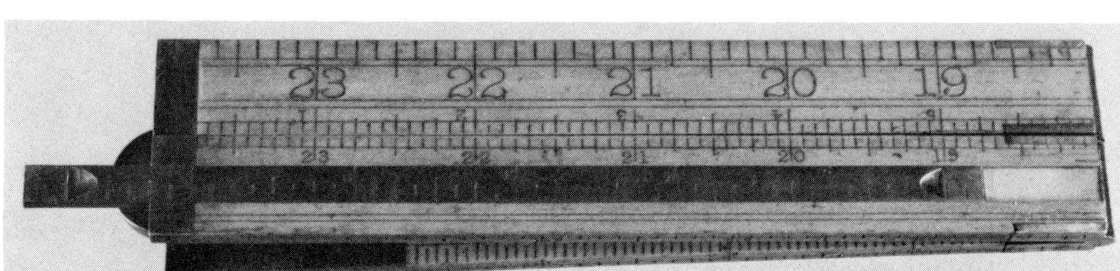

Figure 173: No. 83, 1877—1898 *(Author's Collection)*

Presumably this slide was similar in scale arrangement to that of the carpenters'/engineers' sliding rules, with two-cycle A, B, and C scales, and a single-cycle D scale. It is not known whether the D scale was folded; English makers, from whom this pattern presumably was taken, made both types.

In 1874, the Gunter's scales were deleted from this rule, and the slide graduated in 8ths of inches to function as a simple extension slide (this is the example illustrated here).

All but one or two of the few known examples of this rule exhibit more or less damage to the wood track of the slide near the knuckle joint. This was caused by attempting to open that knuckle joint while the slide is extended in the direction of that joint; when this was attempted, the head stick on the other side of the joint would lever against the underside of the slide and forcibly pry it from the groove, damaging it in the process.

83C Carpenters' Caliper Rule, 2 Foot, 4 Fold

Offered:	1910 thru 1932
Construction:	Arch Joint, Edge Plates, Left Hand Caliper
Material:	Boxwood, Brass Hinges, Caliper & Tips
Width:	$1\frac{3}{8}$ Inch
Graduations:	Body 8ths, 10ths, and 16ths of Inches
	Caliper: 32nds of Inches
Scales:	Drafting

Although officially discontinued after 1932 (Ref. 43), the No. 83C must have continued to be offered on an informal basis (presumably to use up existing stocks) for some years thereafter. The 1942 (Ref. 52) and 1943 (Ref. 53) corrected price lists both refer to it as being listed in the 1941 dealers' catalogue.

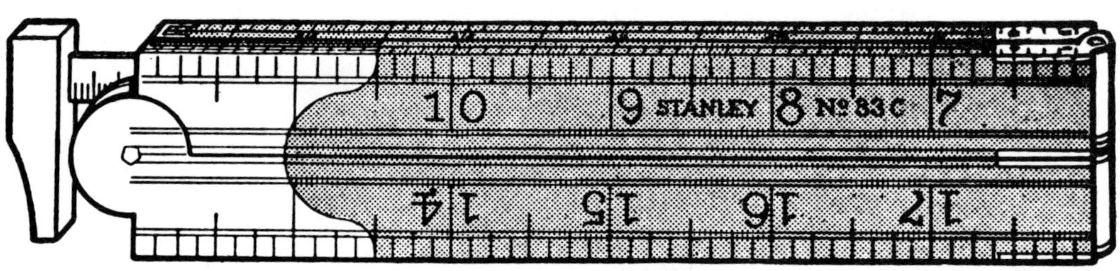

Figure 174: No. 83C, 1919 *(Ref. 33)*

One example has been noted of a rule identical to the No. 83C, but numbered No. 83½. It is not known what relationship this rule bears to the No. 83C. It may have been its predecessor, prior to 1910, but if so, it was only made for a very short time; it was never included in any Stanley catalogue.

84 Carpenters' Rule, 2 Foot, 4 Fold

Offered:	1859 thru 1932
Construction:	Square Joint, Half Bound
Material:	Boxwood, Brass Hinges & Trim
Width:	1 Inch
Graduations:	8ths and 16ths of Inches (1872 and before)
	8ths, 10ths, and 16ths of Inches (1874 thru 1877)
	8ths, 10ths, 12ths, and 16ths of Inches (1879 and after)
Scales:	Drafting

Figure 175: No. 84, 1922–1932 *(Author's Collection)*

85 Carpenters' Rule, 2 Foot, 4 Fold

Offered: 1855 thru 1915
Construction: Square Joint, Edge Plates
10ths Scale on Edge (1862 and before)
Leather Case (1912 and after)
Material: Ivory, Brass or German Silver Hinges & Tips (1859 and before)
Ivory, German Silver Hinges & Tips (1860 and after)
Width: 7/8 Inch
Graduations: 8ths, 10ths, and 16ths of Inches (1872 and before)
8ths, 10ths, 12ths, and 16ths of Inches (1874 and after)

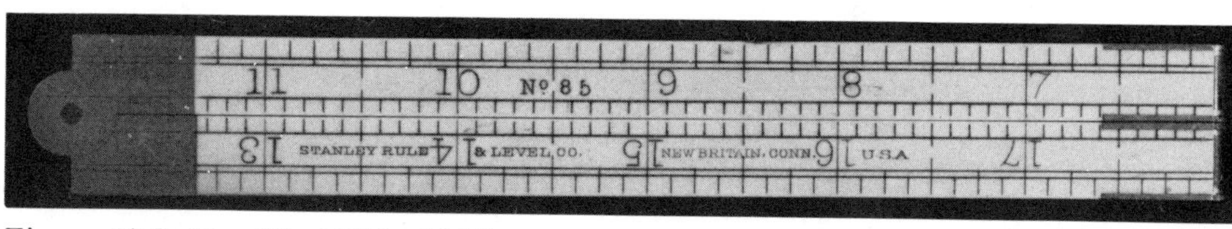

Figure 176: No. 85, 1858–1915 *(James Hill Collection)*

86 Carpenters' Rule, 2 Foot, 4 Fold

Offered: 1855 thru 1922
Construction: Arch Joint, Edge Plates
10ths Scale on Edge (1862 and before)
Leather Case (1912 thru 1917)
Material: Ivory, Brass or German Silver Hinges & Tips (1859 and before)
Ivory, German Silver Hinges & Tips (1860 thru 1917)
Ivory, Nickel Silver Hinges & Tips (1919 and after)
Width: 1 Inch
Graduations: 8ths, 10ths, and 16ths of Inches (1877 and before)
8ths, 10ths, 12ths, and 16ths of Inches (1879 and after)
100ths of a Foot (1865 and after)
Scales: Drafting

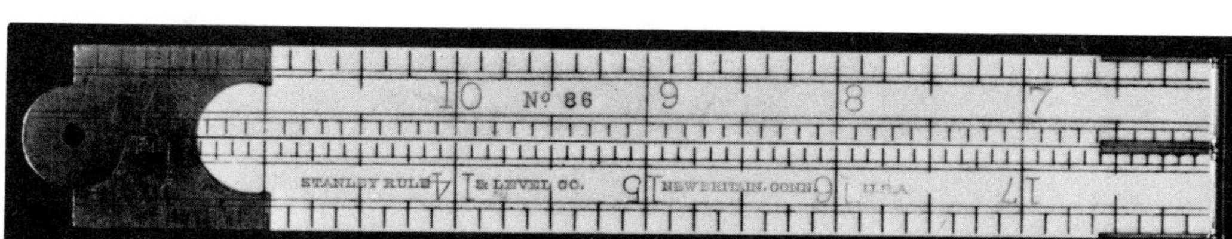

Figure 177: No. 86, 1858–1917 *(John Treggiari Collection)*

86½ Carpenters'/Architects' Rule, 2 Foot, 4 Fold

Offered:	1884 thru 1917
Construction:	Arch Joint, Edge Plates, Inside Beveled Edges Leather Case (1912 and after)
Material:	Ivory, German Silver Hinges & Tips
Width:	1 Inch
Graduations:	8ths, 10ths, 12ths, and 16ths of Inches
Scales:	Drafting

Figure 178: No. 86½, 1884–1917 *(Charles and Walter Jacob Collection)*

87 Carpenters' Rule, 2 Foot, 4 Fold

Offered:	1855 thru 1922
Construction:	Arch Joint, Full Bound Leather Case (1912 thru 1917)
Material:	Ivory, Brass or German Silver Hinges & Trim (1859 and before) Ivory, German Silver Hinges & Trim (1860 thru 1917) Ivory, Nickel Silver Hinges & Trim (1919 and after)
Width:	1 Inch
Graduations:	8ths, 10ths, and 16ths of Inches (1877 and before) 8ths, 10ths, 12ths, and 16ths of Inches (1879 and after)
Scales:	Drafting

Figure 179: No. 87, 1879–1917 *(James Hill Collection)*

88 Carpenters' Rule, 1 Foot, 4 Fold

Offered:	1855 thru 1922
Construction:	Arch Joint, Full Bound Leather Case (1912 thru 1917)
Material:	Ivory, Brass or German Silver Hinges & Trim (1859 and before) Ivory, German Silver Hinges & Trim (1860 thru 1917) Ivory, Nickel Silver Hinges & Trim (1919 and after)
Width:	½ or ⅝ Inch (1860 and before) ⅝ Inch (1862 and after)
Graduations:	8ths and 16ths of Inches

Figure 180: No. 88, 1858–1917 *(Roger K. Smith Collection)*

88½ Carpenters' Rule, 1 Foot, 4 Fold

Offered:	1855 thru 1917
Construction:	Arch Joint, Edge Plates
	Leather Case (1912 and after)
Material:	Ivory, Brass or German Silver Hinges & Tips (1859 and before)
	Ivory, German Silver Hinges & Tips (1860 and after)
Width:	½ or ⅝ Inch (1860 and before)
	⅝ Inch (1862 and after)
Graduations:	8ths and 16ths of Inches

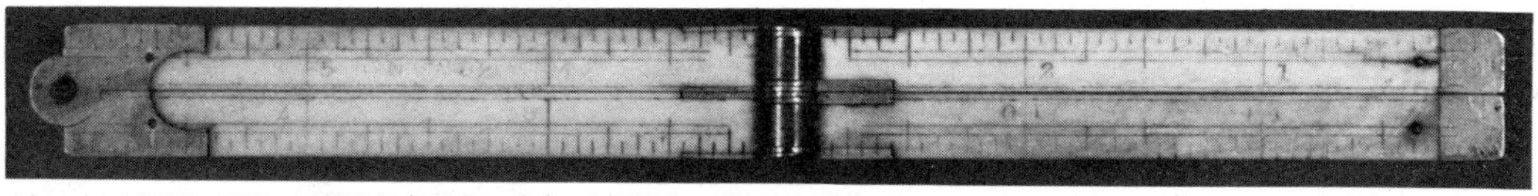

Figure 181: No. 88½ (½ Inch), 1855–1860 *(William Baader Collection)*

89 Carpenters' Rule, 2 Foot, 4 Fold

Offered:	1855 thru 1915
Construction:	Double Arch Joint, Full Bound
	Leather Case (1912 and after)
Material:	Ivory, Brass or German Silver Hinges & Trim (1855)
	Ivory, German Silver Hinges & Trim (1859 and after)
Width:	1 Inch
Graduations:	8ths, 10ths, and 16ths of Inches (1877 and before)
	8ths, 10ths, 12ths, and 16ths of Inches (1879 and after)
Scales:	Drafting

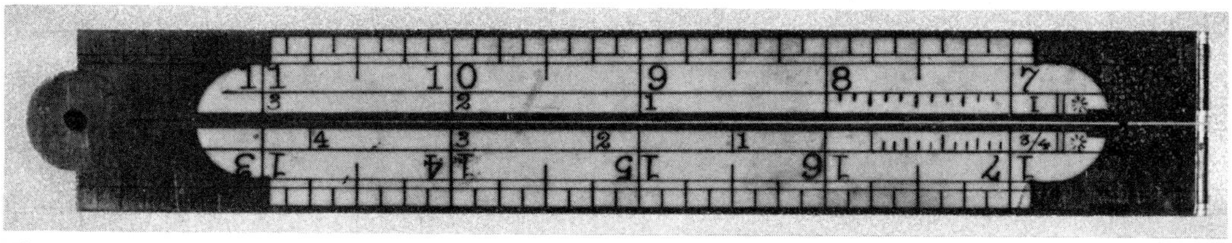

Figure 182: No. 89, 1854–1857 *(Bud Steere Collection)*

90 Carpenters' Rule, 1 Foot, 4 Fold

Offered:	1855 thru 1917
Construction:	Round Joint, Middle Plates
Material:	Ivory, Brass or German Silver Hinges & Tips (1855) Ivory, Brass Hinges & Tips (1859 thru 1907) Ivory, German Silver Hinges & Tips (1909) Ivory, Brass Hinges & Tips (1912 and after)
Width:	"Regular" (5/8 Inch) and "Narrow (1/2 Inch) (1860 and before) Not Known (1862 thru 1907) 1/2 Inch (1909 and after)
Graduations:	8ths and 16ths of Inches

The No. 90 was the simplest and least expensive of Stanley's line of 4 fold ivory rules, and unquestionably the one whose structural details changed most often. At various times it had brass, German silver, or optional trim, switching frequently from one to the other, as the company attempted to keep its cost low. For more than forty years even its width was undefined; this allowed Stanley to use whatever width ivory sticks were surplus from other rules to assemble No. 90's.

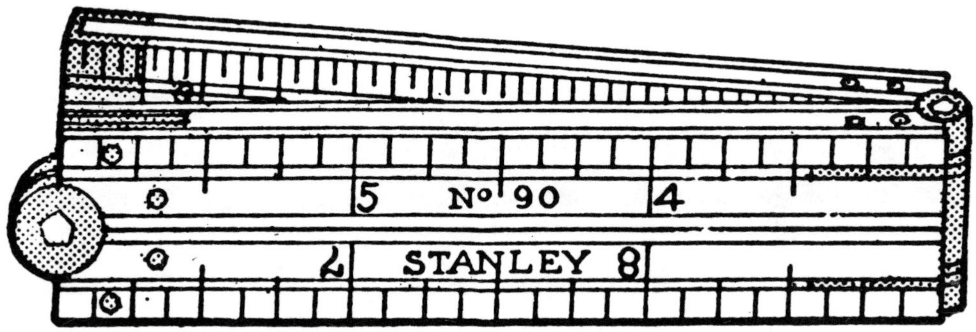

Figure 183: No. 90, 1914 *(Ref. 29)*

The No. 90 was discontinued between 1917 (Ref. 32) and 1919 (Ref. 33). Since this was presumably one of Stanley's better-selling ivory rules, it would not have remained in stock for long after the end of production. This may indicate that production of ivory rules actually stopped in 1917–1918, and those rules offered in 1919 and 1922 were existing stocks only.

90½ Carpenters' Rule, 1 Foot, 4 Fold

Offered:	1855
Construction:	Round Joint, Edge Plates
Material:	Ivory, Brass or German Silver Hinges & Tips
Width:	Not Known
Graduations:	Not Known

91 Carpenters' Rule, 1 Foot, 4 Fold
Offered: 1855 thru 1917
Construction: Square Joint, Edge Plates
Leather Case (1912 and after)
Material: Ivory, Brass or German Silver Hinges & Tips (1859 and before)
Ivory, German Silver Hinges & Tips (1860 and after)
Width: 7/8 Inch (1865 and before)
3/4 Inch (1867 and after)
Graduations: 8ths, 10ths, and 16ths of Inches (1872 and before)
8ths, 10ths, 12ths, and 16ths of Inches (1874 and after)

92 Carpenters' Rule, 1 Foot, 4 Fold
Offered: 1855 thru 1922
Construction: Square Joint, Edge Plates
Material: Ivory, Brass or German Silver Hinges & Tips (1859 and before)
Ivory, German Silver Hinges & Tips (1860 thru 1917)
Ivory, Nickel Silver Hinges & Tips (1919 and after)
Width: 1/2 or 5/8 Inch (1860 and before)
5/8 Inch (1862 and after)
Graduations: 8ths and 16ths of Inches

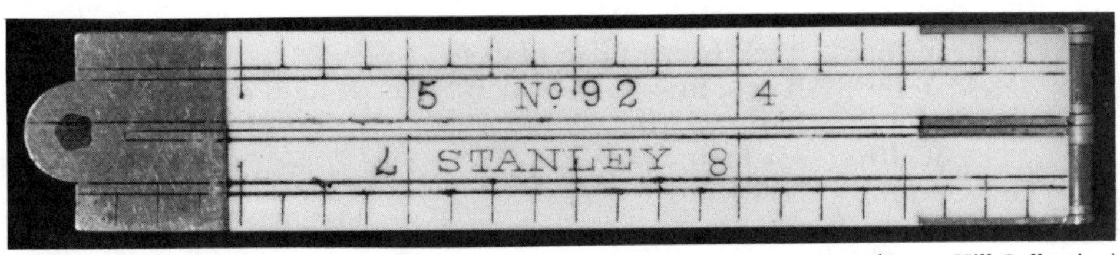

Figure 184: No. 92, Ca. 1900 *(James Hill Collection)*

92½ Carpenters' Rule, 1 Foot, 4 Fold
Offered: 1855 thru 1915
Construction: Square Joint, Middle Plates
Material: Ivory, Brass or German Silver Hinges & Tips (1859 and before)
Ivory, German Silver Hinges & Tips (1860 thru 1909)
Ivory, Brass Hinges & Tips (1912 and after)
Width: 1/2 or 5/8 Inch (1860 and before)
5/8 Inch (1862 and after)
Graduations: 8ths and 16ths of Inches

Figure 185: No. 92½, 1858–1915 *(Charles and Walter Jacob Collection)*

93 Carpenters' Rule, 6 Inch, 2 Fold
Offered: 1855 thru 1872
Construction: Round Joint
Material: Ivory, Brass or German Silver Hinge & Tips (1855)
Ivory, Brass Hinge & Tips (1862 and after)
Width: Not Known
Graduations: 8ths and 16ths of Inches

The No. 93 was not offered in the 1860 price list (Ref. 3). Presumably this was an oversight; the entire catalogue was reset that year, and it would have been easy for a typographical omission like this to slip through.

93½ Carpenters' Rule, 6 Inch, 2 Fold
Offered: 1862 thru 1867
Construction: Round Joint
Material: Ivory, German Silver Hinge & Tips
Width: Not Known
Graduations: 8ths and 16ths of Inches

94 Carpenters' Rule, 2 Foot, 4 Fold
Offered: 1855 thru 1859
Construction: Arch Joint, Edge Plates
Material: Ivory, Brass or German Silver Hinges & Tips (1855)
Ivory, German Silver Hinges & Tips (1859)
Width: 1¼ Inch
Graduations: Not Known

94 Carpenters' Rule, 4 Foot, 4 Fold
Offered: 1879 thru 1943
Construction: Arch Joint, Full Bound
Material: Boxwood, Brass Hinges & Trim
Width: 1½ Inch
Graduations: 8ths and 16ths of Inches

The No. 94 was also referred to in the catalogue as a "carriage makers'" rule, undoubtedly a reference to its length (4 feet) and extra strong construction (it had an arch joint, was full bound, and had triple-plated middle joints). A carriage maker would require his rule to be longer than normal, due to the size of his product, and also stronger, to stand the abuse it would be subjected to on the shop floor.

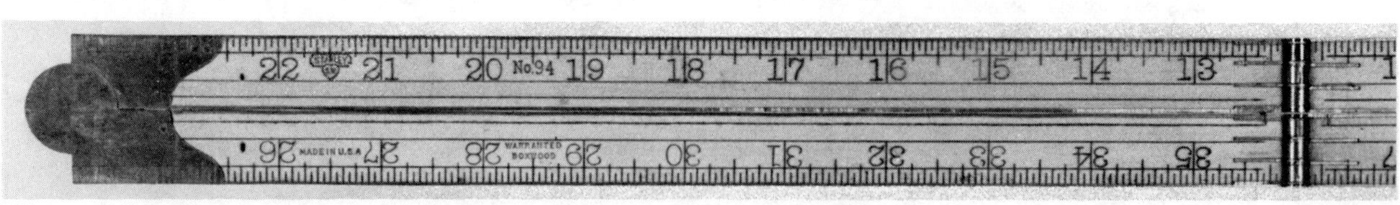

Figure 186: No. 94, 1922–1932 *(Ruben Morrison Collection)*

The No. 94 was dropped some time after 1943 (Ref. 53), but may have actually ceased production some time before that. It is hard to imagine that Stanley had continued making an elaborate rule like this in the face of war production demands. Probably the rules offered in 1942–1943 were existing stocks, and the No. 94 had actually not been made for a number of years.

95 Carpenters' Rule, 2 Foot, 4 Fold

Offered:	1855 thru 1907
Construction:	Arch Joint, Full Bound
Material:	Ivory, Brass or German Silver Hinges & Trim (1855) Ivory, German Silver Hinges & Trim (1859 and after)
Width:	1 3/8 Inch
Graduations:	8ths, 10ths, and 16ths of Inches (1877 and before) 8ths, 10ths, 12ths, and 16ths of Inches (1879 and after)
Scales:	Drafting

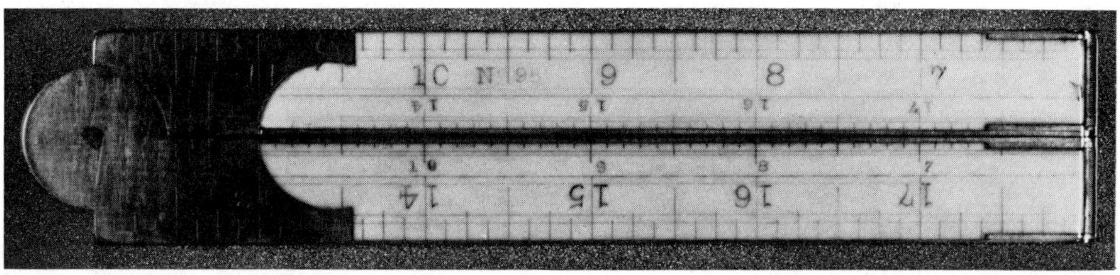

Figure 187: No. 95, 1858–1907 *(Charles and Walter Jacob Collection)*

96 Carpenters' Rule, 2 Foot, 4 Fold

Offered:	1855
Construction:	Double Arch Joint, Edge Plates
Material:	Ivory, German Silver Hinges & Tips
Width:	Not Known, but probably 1¼ Inch
Graduations:	Not Known

97 Carpenters' Rule, 2 Foot, 4 Fold

Offered:	1855 thru 1907
Construction:	Double Arch Joint, Full Bound
Material:	Ivory, Brass or German Silver Hinges & Trim (1855) Ivory, German Silver Hinges & Trim (1859 and after)
Width:	1 3/8 Inch
Graduations:	8ths, 10ths, and 16ths of Inches (1877 and before) 8ths, 10ths, 12ths, and 16ths of Inches (1879 and after)
Scales:	Drafting

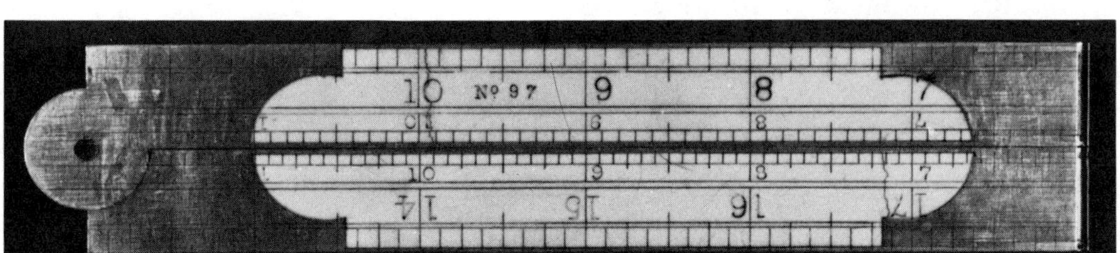

Figure 188: No. 97, 1858–1907 *(Charles and Walter Jacob Collection)*

98 Desk Rule, 1 Foot

Offered: 1867 thru 1877, 1919 thru 1932
Construction: Beveled Edge
Material: Boxwood
Width: ¾ Inch
Graduations: 8ths and 16ths of Inches

Figure 189: No. 98, 1867–1921 *(Author's Collection)*

98M School Rule, 1 Foot

Offered: 1919 thru 1932
Construction: Beveled Edge
Material: Boxwood
Width: ¾ Inch
Graduations: 8ths of Inches
Centimeters/Millimeters

Figure 190: No. 98M, 1927 *(Ref. 40)*

99 School Rule, 1 Foot

Offered: 1867 thru 1877
Construction: Beveled Edge
Material: Boxwood
Width: ¾ Inch
Graduations: 10ths and 16ths of Inches

101 Carpenters' Rule, 2 Foot, 2 Fold

Offered: 1877 thru 1879
Construction: Arch Joint
Material: Boxwood, Brass Hinge & Tips
Width: 1½ Inch
Graduations: Inches/Fractions
Centimeters/Millimeters

The No. 101 was identical to the No. 1, but was graduated on one side in centimeters/millimeters instead of inches/fractions.

Inside-Measuring Calipers

The Nos. 136 and 136½ calipers included a feature which had not previously been offered on Stanley caliper rules. This was the ability to accurately measure inside dimensions and the diamters of holes.

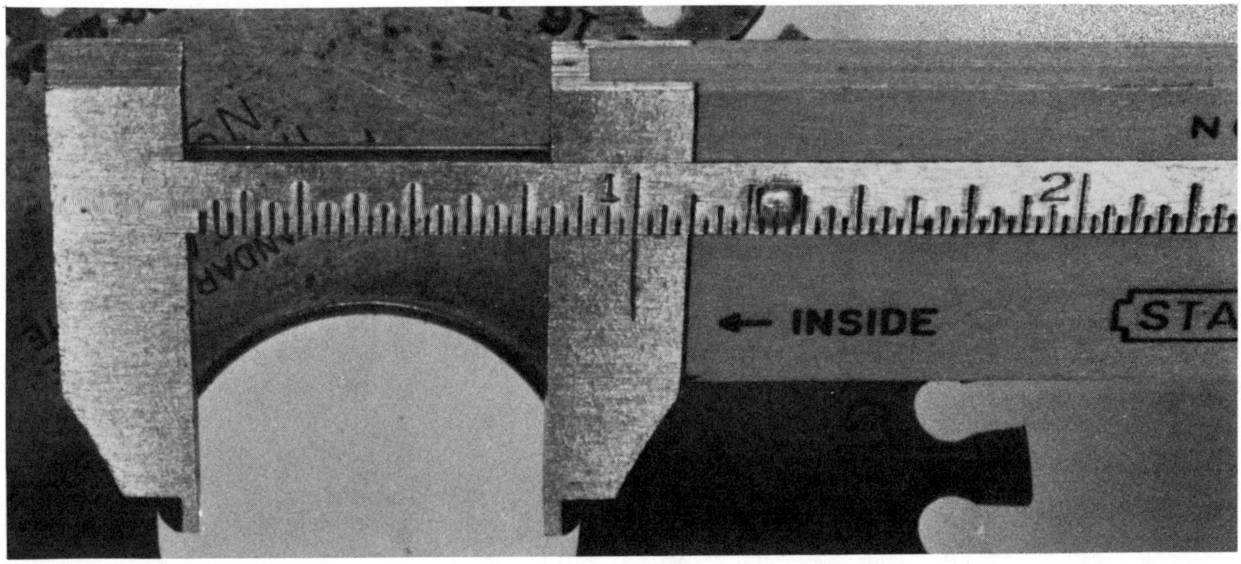

Figure 191: Inside Measurements Using The No. 136 *(Author's Collection)*

This was made possible by a feature which, while common in machinist's hand tools, had not hitherto been applied to boxwood caliper rules. On these two rules the tips of the caliper jaws were milled on the outside into a curved shape of an exact diameter (3/16 inches on the No. 136, 7/32 inches on the No. 136½). On the body of the rule an index line was marked that same distance (3/16 or 7/32 inches) from the edge/line used for reading the scale on the slide when measuring an outside dimension. By using this index line instead, the distance between the outside surfaces of the jaw tips could be accurately read, and the rule be used for inside measurements as well.

136 Caliper Rule, 4 Inch

Offered:	1932 thru 1942, 1947 thru 1958	
Construction:	Caliper, Milled for Inside and Outside Measurements	
Material:	Boxwood, Brass Caliper & Tips	
Width:	1 3/8 Inch	
Graduations:	Body:	16ths of Inches
	Caliper:	32nds of Inches

The Stanley catalogues described the No. 136 as having the capacity to measure hole diameters from 3/16 to 3 1/4 inches, outside diameters of rounds up to 1 7/8 inches (the depth of the jaws), and overall lengths up to 3¼ inches.

The No. 136 was one of the number of rules which was canceled in 1943 to make room for increased war production. The 1943 price list (Ref. 53) did not identify this cancellation as either "permanant" or "temporary," but it must have been the latter; when the first postwar catalogue was issued in 1947 (Ref. 54) the No. 136 was back in the product line.

136R Caliper Rule, 4 Inch
Offered: 1960 thru 1975

Identical to No. 136; a model number change only

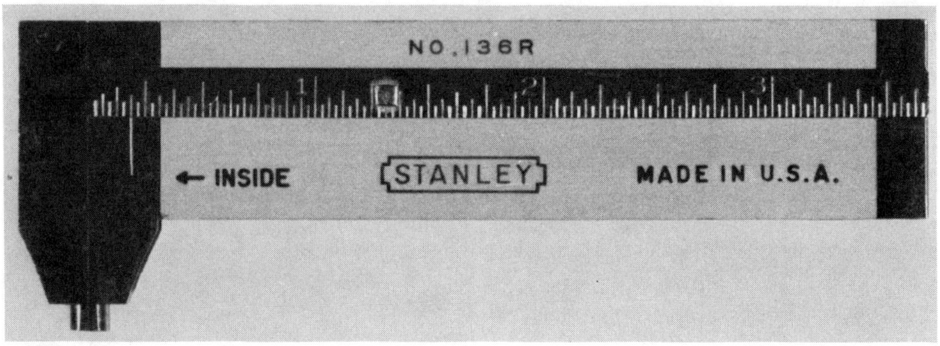

Figure 192: No. 136R, 1960–1975 *(Author's Collection)*

136½ Caliper Rule, 5 7/8 Inch
Offered: 1941 thru 1942, 1947 thru 1975
Construction: Caliper, Milled for Inside and Outside Measurements
Material: Boxwood, Brass Caliper & Tips
Width: 2 1/16 Inch
Graduations: Body: 16ths of Inches
Caliper: 32nds of Inches

The Stanley catalogues described the No. 136½ as having the capacity to measures hole diameters from 7/32 to 5 inches, outside diameters of rounds up to 3 inches (the depth of the jaws), and overall lengths up to 5 inches.

Figure 193: No. 136½, 1941–1975 *(Author's Collection)*

The No. 136½ was one of the number of rules which was canceled in 1943, only two years after it was first offered, to make room for increased war production. The 1943 price list (Ref. 53) did not identify this cancellation as either "permanent" or "temporary," but it must have been the latter; when the first postwar catalogue was issued in 1947 (Ref. 54) the No. 136 was back in the product line.

141 Meter Stick, 1 Meter

Offered: 1925 thru 1943
Material: Maple, Brass Tips
Width: 1 Inch
Graduations: 8ths of Inches
Centimeters/Millimeters

When the No. 141 was first introduced in 1925 (Ref. 38) it did not have a number, but was simply described as a "meter stick, brass tips." The next year this was changed, however, and from 1926 on (Ref. 39) it was listed as the No. 141.

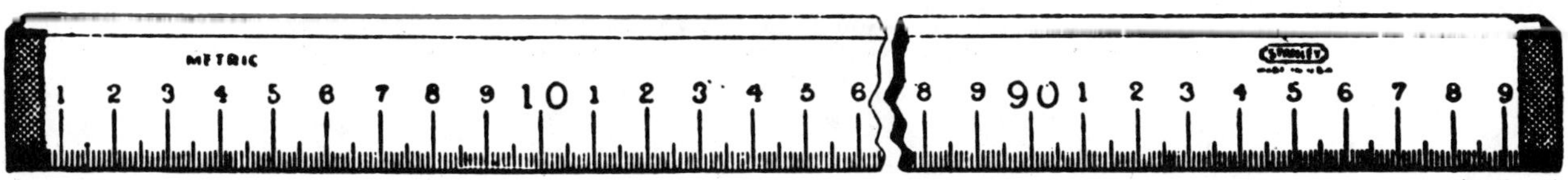

Figure 194: No. 141, 1934 *(Ref. 44)*

This meter stick must have been available on an informal basis for a number of years prior to 1925. At least two examples have been noted bearing the Stanley Rule & Level Co. trademark, evidence of pre-1922 manufacture.

141M Meter Stick, 1 Meter

Offered: 1934 thru 1941
Material: Maple, Brass Tips
Width: 1 Inch
Graduations: Centimeters/Millimeters

The No. 141M was identical to the No. 141, but was graduated with metric divisions on both, instead of only one side.

142 Meter Stick, 1 Meter

Offered: 1925 thru 1941
Material: Maple
Width: 1 Inch
Graduations: 8ths of Inches
Centimeters/Millimeters

The No. 142 was identical to the No. 141, but had plain tips.

When the No. 142 was first introduced in 1925 (Ref. 38) it did not have a number, but was simply described as a "meter stick." The next year this was changed, however, and from 1926 on (Ref. 39) it was listed as the No. 142.

142M Meter Stick, 1 Meter
Offered: 1934 thru 1937
Material: Maple
Width: 1 Inch
Graduations: Centimeters/Millimeters

The No. 142M was identical to the No. 142, but was graduated with metric divisions on both, instead of only one side.

151 Carpenters' Rule, 2 Foot, 4 Fold
Offered: 1877 thru 1879
Construction: Arch Joint, Middle Plates
Material: Boxwood, Brass Hinges & Tips
Width: 1 Inch
Graduations: Inches/Fractions
Centimeters/Millimeters

The No. 151 was identical to the No. 51, but was graduated on the outside in centimeters/millimeters instead of inches/fractions.

Stationers' Goods

For a period of at least three years, from 1862 to 1865, the catalogue included a number of items under the heading Stationers' Goods, as part of the section devoted to measuring instruments. These were items which, while not directly related to the remainder of the product line, were either specialty measuring instruments, such as hatters' rules, or wood items without graduations that might conceivably be sold through stationery stores. This was a period when Stanley had not yet focused their efforts on tool manufacture exclusively, and they were still offering products as disparate as buttons, toy pistols, and even roller skates.

Fourteen of these stationery items have been included here, as being within the purview of this book. Of these, five were plain rulers (used by bookkeepers to rule the lines on the pages of their account books); one was a bevel edge school rule, graduated in 8ths and 16ths; two were drafting scales, one with two scales, the other with sixteen; one was a Gunter's scale (a calculating instrument used by navigators and surveyors), two were the two 1 foot printers' rules, one solid, the other 4 fold, which had been introduced in 1860; the last three were a button gauge (for seamstress' use), a hatters' rule (a hat size gauge), and a watch glass gauge (used by watchmakers). Five others have been omitted as not being of interest: three different sets of chess pieces and checkers, and two different dissected cubes designed for classroom demonstration of cube root, etc.

Only eleven of the fourteen items listed in 1862 (Ref. 5) were still offered in 1865 (Ref. 6): the Nos. 160 through 168 rulers, the No. 170 scholar's rule, the Nos. 172 and 174 drafting scales, the Nos. 186 and 188 printers' rules, and the No. 212 Hatters' Rule. The other three, the No. 176 Gunter's scale, the No. 210 button gauge, and the No. 214 watch glass gauge, had all been eliminated.

In 1867 (Ref. 7) the Stationers' Goods section, and all of the items remaining in it, was removed from the catalogue, never to reappear. This was apparently caused by the concentration of Stanley's efforts on the manufacture of woodworking tools (two years later they were to purchase Bailey, Chany, & Co. and embark upon plane manufacture), and the resultant elimination of extraneous products. The only one to ever reappear was the No. 210 button gauge. Renumbered as the No. 23, and accompanied by its ivory equivalent, the No. 24, it was offered one more time, in 1867 (Ref. 7), before vanishing forever.

160 Plain Ruler, 12 to 18 Inch

Offered: 1862 thru 1865
Construction: Flat, Bevel Edge, Plain Finish
Material: Mahogany
Width: Unknown
Graduations: None

The No. 160 was one of a group of five plain rulers offered under the heading of "Stationers' Goods" in 1862 and 1865. Rulers of this type were not graduated, being intended for use by bookkeepers to rule the lines in their account books.

The No. 160 was intended for retail sale by stationers, and was only offered in wholesale lots of 12 in assorted lengths. In 1862 (Ref. 5) this assortment was of lengths from 12 to 15 inches; later, in 1865 (Ref. 6), a choice of assortments was offered: from 12 to 15 inch, or from 12 to 18 inch.

161 Carpenters' Rule, 2 Foot, 4 Fold

Offered: 1877 thru 1879
Construction: Square Joint, Middle Plates
Material: Boxwood, Brass Hinges & Tips
Width: 1 Inch
Graduations: Inches/Fractions
Centimeters/Millimeters

The No. 161 was identical to the No. 61, but was graduated on the outside in centimeters/millimeters instead of inches/fractions.

162 Plain Ruler, 12 to 18 Inch

Offered: 1862 thru 1865
Construction: Flat, Bevel Edge, French Polish
Material: Mahogany
Width: Unknown
Graduations: None

The No. 162 was one of a group of five plain rulers offered under the heading of "Stationers' Goods" in 1862 and 1865. Rulers of this type were not graduated, being intended for use by bookkeepers to rule the lines in their account books.

The No. 162 was intended for retail sale by stationers, and was only offered in wholesale lots of 12 in assorted lengths. In 1862 (Ref. 5) this assortment was of lengths from 12 to 15 inches; later, in 1865 (Ref. 6), a choice of assortments was offered: from 12 to 15 inch, or from 12 to 18 inch.

163 Carpenters' Rule, 2 Foot, 4 Fold

Offered: 1941 thru 1943
Construction: Round Joint, Middle Plates
Material: Wood, Steel Hinges & Tips
Width: 1 Inch
Graduations: 8ths, 10ths, 12ths, and 16ths of Inches
Scales: Drafting

The No. 163 was an extremely cheaply made rule listed in the catalogue from 1941 to 1943. This rule was made of painted or varnished softwood (examples of both have been encountered), with joints and tips of steel. The graduations were printed on.

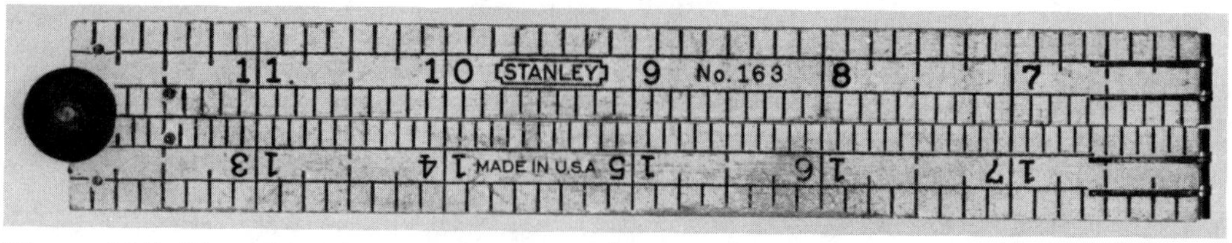

Figure 195: No. 163, 1941–1943 *(Author's Collection)*

This rule was probably a wartime expedient, and most likely was in production longer than the catalogue dates indicate, possibly as late as 1945 or 1946.

164 Plain Ruler, 12 to 18 Inch

Offered: 1862 thru 1865
Construction: Octagonal, French Polish
Material: Mahogany
Width: Unknown
Graduations: Unknown

The No. 164 was one of a group of five plain rulers offered under the heading of "Stationers' Goods" in 1862 and 1865. Rulers of this type were not graduated, being intended for use by bookkeepers to rule the lines in their account books. Octagonal rules were particularly favored for this purpose, as they could be "rolled" down the page, one face at a time, to automatically space the lines uniformly and keep them parallel.

The No. 164 was intended for retail sale by stationers, and was only offered in wholesale lots of 12 in assorted lengths. In 1862 (Ref. 5) this assortment was of lengths from 12 to 15 inches; later, in 1865 (Ref. 6), a choice of assortments was offered: from 12 to 15 inch, or from 12 to 18 inch.

165 Carpenters' Rule, 1 Foot, 4 Fold

Offered: 1877 thru 1879
Construction: Square Joint, Middle Plates
Material: Boxwood, Brass Hinges & Tips
Width: 5/8 Inch
Graduations: Inches/Fractions
Centimeters/Millimeters

The No. 165 was identical to the No. 65, but was graduated on the outside in centimeters/millimeters instead of inches/fractions.

166 Plain Ruler, 12 to 18 Inch

Offered: 1862 thru 1865
Construction: Flat, Bevel Edge, French Polish
Material: Rosewood
Width: Unknown
Graduations: None

The No. 166 was one of a group of five plain rulers offered under the heading of "Stationers' Goods" in 1862 and 1865. Rulers of this type were not graduated, being intended for use by bookkeepers to rule the lines in their account books.

The No. 166 was intended for retail sale by stationers, and was only offered in wholesale lots of 12 in assorted lengths. In 1862 (Ref. 5) this assortment was of lengths from 12 to 15 inches; later, in 1865 (Ref. 6), a choice of assortments was offered: from 12 to 15 inch, or from 12 to 18 inch.

168 Plain Ruler, 12 to 18 Inch

Offered: 1862 thru 1865
Construction: Round, French Polish
Material: Rosewood
Width: Unknown
Graduations: None

The No. 168 was one of a group of five plain rulers offered under the heading of "Stationers' Goods" in 1862 and 1865. Rulers of this type were not graduated, being intended for use by bookkeepers to rule the lines in their account books. Round rules were favored for this purpose, as they could be rolled down the page to automatically keep the lines parallel.

The No. 168 was intended for retail sale by stationers, and was only offered in wholesale lots of 12 in assorted lengths. In 1862 (Ref. 5) this assortment was of lengths from 12 to 15 inches; later, in 1865 (Ref. 6), a choice of assortments was offered: from 12 to 15 inch, or from 12 to 18 inch.

170 Scholars' Rule, 1 Foot

Offered: 1862 thru 1865
Construction: Bevel Edge, "Extra" Finish
Material: Boxwood
Width: Unknown
Graduations: 8ths and 16ths of Inches

170BE Carpenters' ("Blindman's") Rule, 3 Foot, 4 Fold

Offered: 1929 thru 1938
Construction: Square Joint, Middle Plates
Extra Large & Heavy Figures, and Extra Heavy Graduations
Left-to-Right ("English") Graduations
Material: Boxwood, Brass Hinges & Tips
Width: 1 3/8 Inch
Graduations: 8ths and 16ths of Inches

Figure 196: No. 170BE, 1929–1932 *(Charles and Walter Jacob Collection)*

172 Drafting Scale, 1 Foot

Offered: 1862 thru 1865

Construction: Unknown

Material: Boxwood

Width: Unknown

Graduations: Unknown

Scales: Drafting (12ths and 24ths of Inches)

173 Carpenters' Rule, 2 Foot, 4 Fold

Offered: 1877 thru 1879

Construction: Arch Joint, Middle Plates

Material: Boxwood, Brass Hinges & Tips

Width: $1\frac{3}{8}$ Inch

Graduations: Inches/Fractions
Centimeters/Millimeters

The No. 173 was identical to the No. 73, but was graduated on the outside in centimeters/millimeters instead of inches/fractions.

173E Carpenters' Rule, 3 Foot, 4 Fold

Offered: 1929

Construction: Arch Joint, Middle Plates
Left-to-Right ("English") Markings

Material: Boxwood, Brass Hinges & Tips

Width: $1\frac{3}{8}$ Inch

Graduations: 8ths, 10ths, 12ths, and 16ths of Inches

The No. 173E was another of the rules with left to right graduations introduced by Stanley during the 1920's. It was introduced in 1929 (Ref. 41), along with two other similar rules (the Nos. $73\frac{1}{4}$ and $173\frac{1}{4}$E). All three were removed from the catalogue the following year (Ref. 42), probably due to the onset of the Depression of 1929–1939.

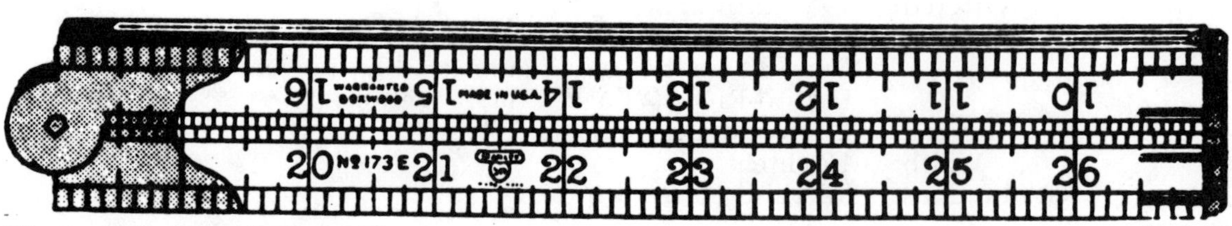

Figure 197: No. 173E, 1929 *(Ref. 41)*

$173\frac{1}{4}$E Carpenters' Rule, 3 Foot, 4 Fold

Offered: 1929

Construction: Arch Joint, Middle Plates
Left-to-Right ("English") Markings

Material: Boxwood, Brass Hinges & Tips

Width: $1\frac{1}{2}$ Inch

Graduations: 8ths, 10ths, 12ths, and 16ths of Inches

The No. 173¼E was another of the rules with left to right graduations introduced by Stanley during the 1920's. It was introduced in 1929 (Ref. 41), along with two other similar rules (the Nos. 73¼ and 173E). All three were removed from the catalogue the following year (Ref. 42), probably due to the onset of the Depression of 1929–1939.

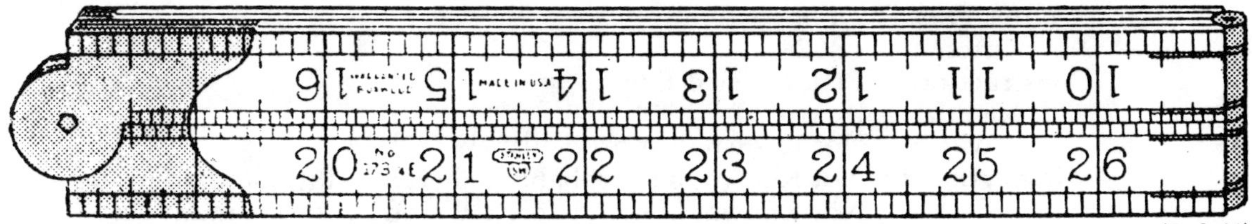

(Ref. 41)

Figure 198: No. 173¼E, 1929

174 Architects' Drafting Scale

Offered:	1862 thru 1865
Construction:	Unknown
Material:	Boxwood
Width:	Unknown
Graduations:	Unknown
Scales:	Drafting; Scale(s) for 1/8, 1/4, 3/8, 1/2, 5/8, 3/4, 7/8, 1, 1¼, 1½, 1¾, 2, 2¼, 2½, 2¾, and 3 Inches to the Foot

The Gunter's Scale

The Gunter's scale is a large rule marked with elaborate scales, which could be used in conjunction with a pair of dividers to solve complex problems in trigonometry and spherical geometry. Invented in 1620–1626 by Edmond Gunter of England, this tool was used up until the end of the nineteenth century by mathematicians and surveyors, and, most extensively, by navigators (seamen referred to it colloquially as a "Gunter").

In its most common form the Gunter's scale was a flat rule, typically 1½ to 2 inches wide, ¼ to 3/8 inch thick, and either 1 or (more commonly) 2 feet long. One edge was usually beveled, and the more elaborate specimens would have the other edge or the tips brass bound.

These rules were almost always graduated to a standard pattern, as follows: On the left half of the front surface was marked an 11 inch scale diagonally divided at one end into 100ths of a half inch and at the other into 100ths of an inch. On the right half of the front surface were marked the 11 so-called "plain" or "natural" lines of numbers, 11 inch scales representing various functions from trigonometry and spherical geometry (sines, tangents, chords, rhumbs, etc.). The upper (unbeveled) edge of the front surface was graduated from right to left in 10ths of inches. On the back surface were marked the 7 "logarithmic" or "artificial" lines of numbers, 24 inch scales representing the logarithms of functions ('versed sines, sines of rhumbs, tangents of rhumbs, etc.), and an 8th scale representing the meridian line on Mercator's chart.

Set flush into the surfaces of the scale were 10 to 15 tiny brass plugs, each with a small dimple punched in its face. These plugs were located at the origins of the various scales, or at key values in the graduations, and were placed there to provide a good location for the tip of a pair of dividers when setting them or reading their setting. They also served to protect the surface of the wood from damage due to repeated piercing by those tips.

All calculations were performed on the Gunter's scale geometrically, physical distances being transferred from one scale to another with the aforementioned dividers. Addition, subtraction, and the taking of trionometric functions were performed on the plain scales; multiplication, division, and powers & roots on the logarithmic ones.

The Gunter's scale eventually fell into disuse toward the end of the nineteenth century. Its method of operation admitted of significant errors, and it was eventually superseded by the logarithmic slide rule, as that device became more sophisticad and accurate.

176 Gunter's Scale, 2 Foot
Offered: 1862
Construction: Flat, Beveled Edge
Material: Boxwood
Width: Not Known, but probably 1½ Inch
Graduations: Not Known, but probably adhered to the standard pattern
Scales: Not Known, but probably adhered to the standard pattern

Printers' Rules

Printers' rules were specially graduated rules used by printers in composing and setting type. Printers were not so much concerned with inches as with number of lines of type, and they used rules graduated in those units when laying out a page or estimating column length, etc.

The traditional form of printers' rule was a ¾ inch square stick, 1 foot long, marked with eight scales, one on each edge of each surface. Seven of these scales would be for the seven most commonly used type sizes: Nonpareil, Minion, Brevier, Bourgeois, Long Primer, Small Pica, and Pica (see PRINTERS' MEASURE, Chapter 3). The eighth scale was usually graduated in Agates, a unit used by printers for measuring columns for gross estimating purposes.

Stanley offered two printers' rules during the period from 1860 (Ref. 4) to 1865 (Ref. 6): a solid, square rule of satinwood, and a flat, 4 fold one of boxwood, both 1 foot long. The square rule almost certainly conformed to the pattern described above, but there is no such basis for forming any surmise as to the other. These rules were first listed under Miscellaneous Rules in the 1860 catalogue, but after Stationers' Goods were introduced in 1862 (Ref. 5), they were transferred to that section, being outside of the normal line of rules for carpenters and mechanics. Both were discontinued in 1867 (Ref. 7) when that section was dropped as part of the company's effort to concentrate exclusively on the manufacture of woodworking tools.

186 Printers' Rule, 1 Foot
Offered: 1860 thru 1865
Construction: Unknown
Material: Satin Wood
Width: Unknown, but probably ¾ Inch
Graduations: None
Scales: Various Type Scales (Nonpariel, Pica Brevier, etc.)
Agate

This rule was never illustrated in any catalogue, and no identifiable example is known to exist. The only available information is the fragmentary facts which can be elicited from the printed description. It is probable that the No. 186 conformed fairly closely to the standard pattern of square printers' rule (see PRINTERS' RULES, above), but without a marked example to examine we cannot be sure.

188 Printers' Rule, 1 Foot, 4 Fold

Offered:	1860 thru 1865
Construction:	Unknown
Material:	Boxwood, probably Brass Trim
Width:	Unknown, but probably 5/8 Inch
Graduations:	None
Scales:	Various Type Scales (Nonpareil, Pica, Brevier, etc.) Agate

This rule was never illustrated in any catalogue, and no identifiable example is known to exist. The only available information is the fragmentary facts which can be elicited from the printed description. It can be presumed that it was one of the standard 4 fold rules, such as the Nos. 55, 56 etc., with printers' graduations in place of the regular ones, and was probably 5/8 inch wide, to allow room for as many scales as possible. The selection and arrangement of the scales, however, is totally unknown, and can only be guessed at.

210 Caliper Button Gauge

Offered:	1862	
Construction:	Caliper	
Material:	Boxwood	
Width:	Unknown	
Length:	Unknown	
Graduations:	Body:	Not Known
	Caliper:	40ths of Inches

211C Yard Stick, 3 Foot

Offered:	1960
Construction:	Varnished
Material:	Basswood
Width:	Unknown
Graduations:	Unknown

Hatters' Rules

A hatters' rule, also known as a hat size gauge, is a short nonfolding rule with an extension slide, used by hatters to measure the inside diameter of customers' old hats preparatory to fitting them with new ones. Usually 5 inches long, and extending to almost 9 inches, these rules were graduated in 8ths of inches, in both directions, on the front surface and from left to right on the slide, and had a table on the back relating hat size, head circumference, and hat circumference (around the outside of the band).

Typically these rules would have a captive slide (which could not be fully removed from the body), with the scale figured to be self indicating. The graduations on the slide would begin at five at the tip, and progress to nine near the right end; as the slide was extended to the left, the graduations and figures on the slide at the point where it left the body would give the total overall length.

Stanley offered a hatters' rule, the No. 212, as part of the group of Stationers' Supplies which were listed in the 1862 (Ref. 5) and 1865 (Ref. 6) catalogues. When this class of item was dropped in 1867, as part of the company's effort to concentrate on woodworking tools, the No. 212 was discontinued as well, after being offered for only 4 years.

212 Hatters' Rule, 5 Inch
Offered: 1862 thru 1865
Construction: Full-Length Slide
Material: Boxwood, Brass Tips
Width: ¾ Inch
Graduations: 8ths of Inches
Other Markings: Table of Hat and Head Sizes

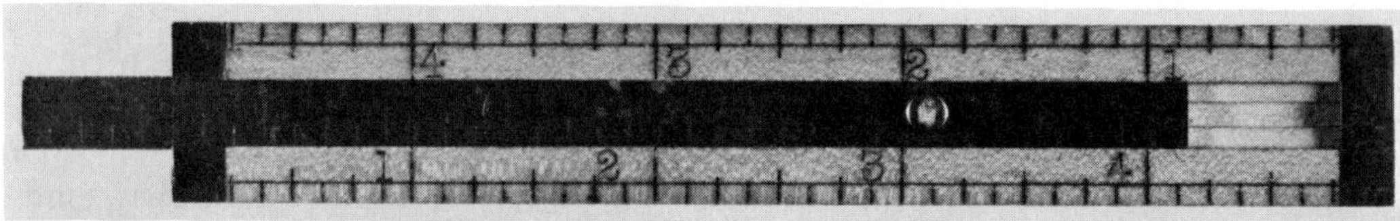

Figure 199a: No. 212, 1862–1865 *(Author's Collection)*

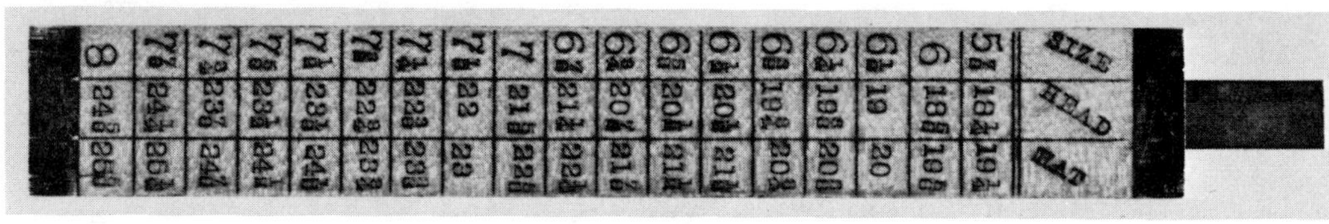

Figure 199b: No. 212, Head/Hat Size Tables *(Author's Collection)*

214 Watch Glass Gauge, Length Unknown
Offered: 1862
Construction: Unknown
Material: Unknown
Width: Unknown
Graduations: Unknown

Nothing is known about Stanley's No. 214 watch glass gauge. There was no standard pattern or system of graduations for these devices, each maker of watches having his own unique set of standard sizes for the glasses.

214 Yard Stick, 3 Foot
Offered: 1941 thru 1942
Construction: Not Known
Material: Not Known
Width: Not Known
Graduations: 8ths of Inches
Fractions of a Yard

214T Yard Stick, 3 Foot
Offered: 1940 thru 1943
Construction: Metal edge, ¼ Inch eyelet in one end
Material: Wood, Brass Tips
Width: 1⅛ Inch
Graduations: 8ths of Inches
Fractions of a Yard

Figure 200: No. 214T, 1940–1943 *(Author's Collection)*

Extension Sticks

A carpenters' extension stick is actually a pair of sticks which slide outward along one another and are used to measure or gauge a surface-to-surface inside dimension.

In the period before arithmetic skills were widespread, such sticks would not have been graduated. The user would have simply captured the dimension by scribing a line on one stick where the other one overlapped; later, back at the shop, by aligning the sticks to that mark the dimension could be recaptured and used.

With the spread of elementary education these sticks became more sophisticated. They were graduated (or made from a pair of yardsticks, etc.), so that they could be used to measure as well as gauge. Clips were added to keep them aligned when sliding, and a clamping mechanism to hold a setting once it had been found.

This is the form in which Stanley began making extension sticks (then called extension "rules") in 1898. The two sections were made of maple, 1 inch by ¼ inch, and had brass tips. They were held together by a pair of cast brass brackets, one of which was equipped with a clamping thumbscrew, and were prevented from sliding completely apart by a round head screw driven into the tip of the back section.

They were offered in four nominal lengths: 2 to 4 foot, 3 to 6 foot, 4 to 8 foot, and 5 to 10 foot, but in each case the actual minimum and maximum length were a little longer than the figure given. The 2 to 4 foot stick, for instance, was 2 feet, 2 inches long when fully collapsed, and 4 feet, 1½ inches long when fully extended.

These sticks were graduated in 8ths of inches, in such a way as to be direct reading. The front section was graduated from right to left, beginning at zero; the back section, which slid out to the left from behind it, was graduated from left to right, beginning at the collapsed length of the stick (on the 2 to 4 foot stick, for instance, the scale on the back section began at 2 feet, 2 inches). Thus as the stick was extended, the graduation on the back section directly opposite the left end of the front section represented the overall length of the stick at that instant. They were not graduated on the back.

In 1907 (Ref. 23) a 6 to 12 foot stick was added to the line, bringing to five the number of different lengths available. In 1909 (Ref. 25) these rules were finally assigned numbers (until this time they had been simply referred to as "extension rules" of a given length). The numbers assigned were derived from the length of the particular stick; the 2

Figure 201: Stanley Extension Stick, 2 foot to 4 foot, 1898 *(Ref. 17)*

to 4 foot stick becoming the No. 240, the 3 to 6 foot the No. 360, and the others the Nos. 480, 510, and 612 as appropriate.

In 1929, Stanley changed their terminology with reference to these devices and began to call them extension "sticks" instead of extension "rules," as theretofor. It is not known why this was done; perhaps they were simply bowing to accepted usage, finally employing the expression by which they had become known to the trade.

Beginning in 1934, the extension sticks began to be eliminated from the line, supplanted by the Pull-Push steel rule and the Zig-Zag rule with an extension slide. The No. 360 (last offered in 1932) was the first to go, followed by the Nos. 240, 480, and 612, all discontinued after 1943 (Ref. 53). The only stick to survive the war was the No. 510; it had been discontinued after 1943, along with all the others, but was resurrected in 1947 (Ref. 54), and was listed for another twenty-one years, until 1968 (Ref. 66).

240 Extension Stick, 2 Foot to 4 Foot
Offered: 1898 thru 1943
Material: Maple, Brass Clamps & Tips
Width: 1 Inch
Graduations: 8ths of Inches

The No. 240 was identical to all of the other Stanley extension sticks (the Nos. 360, 480, 510, and 612) except for extension range. Extending from 2 feet, 2 inches long to 4 feet, 1½ inches long, this was the shortest member of this group of rules, and, judging from the number of surviving examples, must have been one of the more popular as well. Perhaps this was due to its compact length when fully retracted; a stick 26 inches long could be carried and stored much more easily than a longer one of 3 feet or more.

299 Cotton Staple Gauge, 3 Inch
Offered: 1929
Construction: Beveled Edges
Material: Boxwood
Width: 1 Inch
Graduations: 16ths of Inches
Millimeters

The cotton staple gauge was a small rule used in the textile industry to measure thread size and number of threads per inch in fabric.

Figure 202: No. 299, 1929 *(James Hill Collection)*

These gauges may have been made for a long time prior to their one appearance in the 1929 catalogue (Ref. 41). At least two examples have been found made of ivory. Since Stanley stopped making ivory rules sometime around 1918–1920, this would indicate that this gauge was being sold informally at least 10 years prior to 1929.

360 Extension Stick, 3 Foot to 6 Foot
Offered: 1898 thru 1932
Material: Maple, Brass Clamps & Tips
Width: 1 Inch
Graduations: 8ths of Inches

The No. 360 was identical to all of the other Stanley extension sticks (the Nos. 240, 480, 510, and 612) except for extension range: 3 feet, 2 inches to 6 feet, 1½ inches.

Although this rule had been dropped from the catalogue in 1934 (Ref. 44), it may still have been available on an informal basis for some years thereafter. In the 1942 and 1943 supplements to the 1941 dealers' catalogue (Refs. 52 and 53) it is listed as being still available; this could be interpreted as evidence of continued manufacture, or may simply have been an effort to get rid of sticks left over since 1934.

Counter Measures

A counter measure is a very thin form of yardstick designed to be fastened to the surface of a counter so as to prevent its loss and to make measuring simpler. Graduated in fractions of a yard (see THE YARD STICK, earlier in this chapter), these measures were much used in fabric and drapery shops prior to the invention of the rotary fabric gauge in the 1920's.

Stanley made two such measures, the Nos. 450 and 550, both added to the catalogue in 1925. They were of almost identical design, the major difference between them being that the No. 450 was a plain measure, designed to screw onto the surface of the counter, while the No. 550 was stapled to a square stick and could be mounted on the edge.

These counter measures were added to the line as a result of the acquisition by Stanley of the Eagle Square Company in 1916. In addition to its very wide line of steel squares, Eagle Square also made other measuring products, including a steel counter measure.

Figure 203: Eagle Square Counter Measure, Pre-1925 *(Charles and Walter Jacob Collection)*

After continuing this device for a few years under the Eagle Square name, in 1925 (Ref. 38) Stanley redesigned it slightly, and introduced it in two varieties, the Nos. 450 and 550.

These measures were not in the product line for long, however. The rotary fabric gauge was both more accurate and easier to use, and the demand for counter measures was small (although it continued good for wood yardsticks which could be carried around the store). When the Depression of 1929–1939 made reduction of the product line a necessity, these measures were two of those which went; both were eliminated between 1932 (Ref. 43) and 1934 (Ref. 44).

450 Yard Measure, 3 Foot

Offered:	1925 thru 1932
Construction:	Made for attachment to wood counter surface
Material:	Steel, Nickel Plated
Width:	¾ Inch
Thickness:	1/16 Inch
Graduations:	0 thru 9 Inches 8ths of a Yard

The No. 450 counter measure, together with its companion rule, the No. 550 counter measure, are the only steel rules which have been included in this study. The No. 450 was 1/16 inch thick and nickel plated, and its graduations were etched into its surface and filled with black paint. It was graduated from left to right with a hybrid scale where every 8th yard and, additionally, every inch for the first 9 inches, was marked.

(Ref. 40)

Figure 204: No. 450, 1927

This measure was predrilled with holes and supplied with round head nails to fasten it to the counter, and, to prevent catching and tearing of the fabric being measured, its ends were slightly rounded.

As an aid to use, the nail holes were located at the three ¼ yard points on the scale, and at the beginning and ending points as well. Two holes were drilled, side by side, at each point; the round nail heads, protruding above the surface of the rule, allowed the measure to be used by feel alone, if the fabric was obscuring the scale.

480 Extension Stick, 4 Foot to 8 Foot

Offered:	1898 thru 1943
Material:	Maple, Brass Clamps & Tips
Width:	1 Inch
Graduations:	8ths of Inches

The No. 480 was identical to all of the other Stanley extension sticks (the Nos. 240, 360, 510, and 612) except for extension range: 4 feet, 2 inches to 8 feet, 1½ inches.

The illustration of this rule shown in Figure 205 is a classic example of how the most glaring catalogue error can elude the notice of a proofreader for literally years. For thirteen years, from 1925 to 1938, this was the drawing used to depict the No. 240 extension stick in all Stanley catalogues (note the rule number, plainly visible, next to the trademark). However, examination of the graduations clearly demonstrates that this is a No. 480 rule! If this had been a No. 240, the left-most foot figure on the front section would have been a 2, not a 4.

It is not known why this error was allowed to persist this long. Surely someone must have noticed it within a short time of its first appearance in 1925; one would think it would have been quietly corrected by 1926 or 1927, at the latest.

(Ref. 44)

Figure 205: No. "480," 1934

510 Extension Stick, 5 Foot to 10 Foot

Offered: 1898 thru 1936, 1949 thru 1968
Material: Maple, Brass Clamps & Tips (1949 and before)
Maple, Brass-Plated Clamps, Brass Tips (1955 and after)
Width: 1 Inch
Graduations: 8ths of Inches

The No. 510 was identical to all of the other Stanley extension sticks (the Nos. 240, 360, 480, and 612) except for extension range: 5 feet, 2 inches to 10 feet, 1½ inches. This rule was the longest lived of all the extension sticks; it was one of the five sticks originally introduced in 1898, and was the only one returned to the product line after World War II. Except for the period 1937 through 1948, it was in production for fully 70 years.

In 1937 the No. 510 was removed from the catalogue, to be replaced by a similar extension stick, the No. H510 (q.v.). This new extension stick, identical to the No. 510 except for being equipped at one and with a folding hook, was only in the product line for a few years, being discontinued along with all the other remaining extension sticks in 1943. When a 5 to 10 foot extension stick was returned to the catalogue in 1949, it was the No. 510 which was listed, from then until its final cancellation in 1968.

Figure 206a: No. 510, 1949 *(Ref. 55)*

Stanley actually discontinued production of their No. 510 in about 1950–1954. The 5 to 10 foot extension sticks marketed from 1955 thru 1968 were Lufkin No. 7165 extension sticks, marked with the Stanley rule number and trademark (Figure 206b). These, in turn, were discontinued between 1968 (Ref. 66) and 1972 (Ref. 67), finally bringing to an end the line of Stanley extension sticks.

Figure 206b: No. 510, 1955–1968 *(James Hill Collection)*

H510 Extension Stick, 5 Foot to 10 Foot
Offered: 1937 thru 1943

The No. H510 extension stick was identical to the No. 510 which preceded and followed it, except that the No. H510 had a small retracting hook attached to the front surface of the front section at the right hand end (see Fig. 207). This hook, made of flat steel, was to facilitate the use of this stick for outside measurements and for situations where the far edge of the stock was beyond arms' reach. When not needed, it could be folded out of the way, and the H510 used as an ordinary extension stick.

Apparently this special feature was not in high demand; when the decision was made to reintroduce an extension stick after World War II, it was the No. 510 which reappeared in the catalogue, not the No. H510.

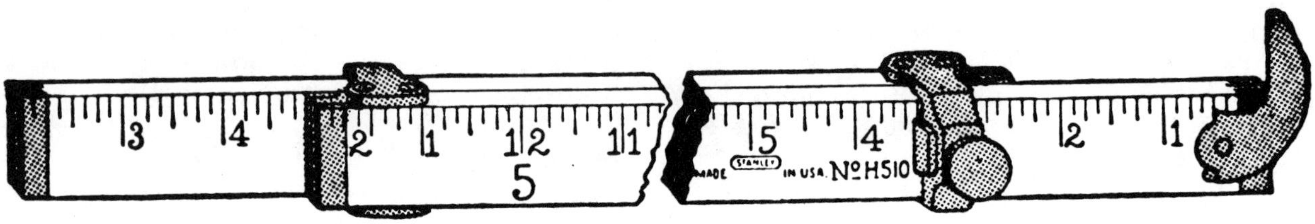

Figure 207: No. H510, 1938 (Ref. 48)

550 Yard Measure, 3 Foot
Offered: 1925 thru 1932
Construction: Made for attachment to wood counter edge
Material: Steel, Nickel Plated
Width: ¾ Inch
Thickness: 1/16 Inch
Graduations: 0 thru 9 Inches
8ths of a Yard

The No. 550 was a variation of the No. 450 counter measure, designed for mounting on the vertical edge of the counter instead of on its surface. This was achieved by mounting the flat measure on a ¾ inch square stick with transverse holes thorugh which screws could be driven into the edge of the counter. This measure was slightly different than the No. 450; instead of rounded ends, it had square ends, allowing it to be rabbeted flush into the surface of the square stick. Also, instead of nails, it was fastened to the stick with staples, leaving a ridge of wire, even easier to feel than a nail head, at each of the ¼ yard points.

Figure 208: No. 550, 1926 (Ref. 89)

612 Extension Stick, 6 Foot to 12 Foot
Offered: 1907 thru 1943
Material: Maple, Brass Clamps & Tips
Width: 1 Inch
Graduations: 8ths of Inches

The No. 612 was identical to all of the other Stanley extension sticks (the Nos. 240, 360, 480, and 510) except for extension range. This stick extended from 6 feet, 2 inches long to 12 feet, 1½ inches long, and was the longest member of this family of extension sticks.

This extreme length must have made the No. 612 sensitive to mishandling and prone to breakage. Most known examples of this rule have repaired breaks in the areas near the brackets on one or the other section. If an extension stick is extended much past about eighty percent of its maximum length, the parts of the sections which still overlap are subject to great strains during handling, and may break. This is particularly true of the long sticks, and probably accounts for the number of broken No. 612's.

CHAPTER 6 Bibliography

Catalogs

1 *Price List of Boxwood and Ivory Rules,* A. Stanley & Co., January 1855 (reprinted by Ken Roberts Publishing Co.)

2 *Price List of Boxwood and Ivory Rules, Levels, Try Squares, Sliding T Bevels, Gauges, &c.,* Stanley Rule and Level Co., February 1859 (reprinted by Ken Roberts Publishing Co.)

3 *Price List of Boxwood and Ivory Rules, Levels, Try Squares, Sliding T Bevels, Gauges, &c.,* Stanley Rule and Level Co., January 1860

4 *Belcher Brothers & Co.'s Price List of Boxwood & Ivory Rules,* Belcher Brothers & Co., 1860 (reprinted by Ken Roberts Publishing Co.)

5 *Price List of U.S. Standard Boxwood and Ivory Rules, Levels, Try Squares, Gauges, Handles, Mallets, Handscrews, &c.,* Stanley Rule and Level Co., July 1862

6 *Price List of U.S. Standard Boxwood and Ivory Rules, Levels, Try Squares, Gauges, Handles, Mallets, Handscrews, &c.,* Stanley Rule and Level Co., 1865

7 *Price List of U.S. Standard Boxwood and Ivory Rules, Levels, Try Squares, Gauges, Handles, Mallets, Handscrews, &c.,* Stanley Rule and Level Co., January 1, 1867 (reprinted by Ken Roberts Publishing Co.)

8 *Price List of U.S. Standard Boxwood and Ivory Rules, Levels, Try Squares, Gauges, Iron and Wood Bench Planes, Mallets, Hand Screws, Spoke Shaves, &c.,* Stanley Rule and Level Co., January 1870 (reprinted by Ken Roberts Publishing Co.)

9 *Price List of U.S. Standard Boxwood and Ivory Rules, Levels, Try Squares, Gauges, Iron and Wood Bench Planes, Mallets, Hand Screws, Spoke Shaves, Screw Drivers, Etc.,* Stanley Rule and Level Co., January 1872 (reprinted by Ken Roberts Publishing Co.)

10 *Price List of U.S. Standard Boxwood and Ivory Rules, Levels, Try Squares, Gauges, Mallets, Screw Drivers, Hand Screws, Iron and Wood Planes, Spoke Shaves, Etc.,* Stanley Rule and Level Co., January 1874 (reprinted by Ken Roberts Publishing Co.)

11 *Price List of U.S. Standard Boxwood and Ivory Rules, Plumbs and Levels, Try Squares, Bevels, Gauges, Mallets, Handles, Awl Hafts, Screw Drivers, Hand Screws, Iron and Wood Planes, Spoke Shaves, Etc.,* Stanley Rule and Level Co., January 1877 (reprinted by Ken Roberts Publishing Co.)

12 *Price List of U.S. Standard Boxwood and Ivory Rules, Plumbs and Levels, Try Squares, Bevels, Gauges, Mallets, Iron and Wood Adjustable Planes, Spoke Shaves, Screw Drivers, Awl Hafts, Handles, Etc.,* Stanley Rule and Level Co., January 1879 (reprinted by Ken Roberts Publishing Co.)

13 *Price List of U.S. Standard Boxwood and Ivory Rules, Plumbs and Levels, Try Squares, Bevels, Gauges, Mallets, Iron and Wood Adjustable Planes, Spoke Shaves, Screw Drivers, Awl Hafts, Handles, Etc.*, Stanley Rule and Level Co., January 1884 (reprinted by Ken Roberts Publishing Co.)

14 *Price List of U.S. Standard Boxwood and Ivory Rules, Plumbs and Levels, Try Squares, Bevels, Gauges, Mallets, Iron and Wood Adjustable Planes, Spoke Shaves, Screw Drivers, Awl Hafts, Handles, Etc.*, Stanley Rule and Level Co., January 1888 (reprinted by Ken Roberts Publishing Co.)

15 *Price List of U.S. Standard Boxwood and Ivory Rules, Plumbs and Levels, Try Squares, Bevels, Gauges, Mallets, Iron and Wood Adjustable Planes, Spoke Shaves, Screw Drivers, Awl Hafts, Handles, Etc.*, Stanley Rule and Level Co., January 1892 (reprinted by H.C. Maddocks, Jr.)

16 *Catalogue of Measuring Rules, Tapes, Straight Edges, and Steel Band Chains; Spirit Levels, &c.*, John Rabone & Sons, Birmingham, England, July 1892 (reprinted by Ken Roberts Publishing Co.)

17 *Price List of U.S. Standard Boxwood and Ivory Rules, Plumbs and Levels, Try Squares, Bevels, Gauges, Mallets, Iron and Wood Adjustable Planes, Spoke Shaves, Screw Drivers, Awl Hafts, Handles, &c.*, Stanley Rule and Level Co., January 1898 (reprinted by Ken Roberts Publishing Co.)

18 *A Mechanics' Pocket Book of U.S. Standard Rules, Plumbs and Levels, Iron and Wooden Planes, Try Squares and Bevels, and Miscellaneous Improved Woodworkers Tools, Catalogue No. 26,* Stanley Rule and Level Co., August 1900

19 *Illustrated Price List of Rules, Spirit Levels, Planes and Tools, &c.,* Edward Preston & Sons, Ltd., Birmangham, England, July, 1901 (reprinted by Ken Roberts Publishing Co.)

20 *Stanley Rule and Level Co., Catalogue No. 28,* Stanley Rule and Level Co., January 1902

21 *Wood-Workers' Tools, Catalogue No. 21,* T.B. Rayl Co., Ca. 1905 (reprinted by Roger K. Smith)

22 *A Mechanics' Reference Book Containing U.S. Standard Rules, Plumbs and Levels, Iron and Wood Planes, Try Squares and Bevels, Bit Braces, Mitre Boxes, Screw Drivers, and Miscellaneous Improved Woodworkers Tools, Catalogue No. 34,* Stanley Rule and Level Co., 1905

23 *A Mechanics' Reference Book Containing U.S. Standard Rules, Plumbs and Levels, Iron and Wood Planes, Try Squares and Bevels, Bit Braces, Mitre Boxes, Screw Drivers, and Miscellaneous Improved Woodworkers Tools, Catalogue No. 34* (2nd Edition), Stanley Rule and Level Co., 1907

24 *Trade Price List, Catalogue No. 39,* Stanley Rule and Level Co., 1908

25 *Carpenters & Mechanics Tools, Catalogue No. 102,* Stanley Rule and Level Co., 1909 (reprinted by Roger K. Smith)

26 *Stanley Tools, Catalogue No. 110,* Stanley Rule and Level Co., 1910

27 *Stanley Tools, Catalogue No. 34,* Stanley Rule & Level Co., 1912

28 *Complete Woodworkers' Tool Catalogue,* Orr & Lockett Hardware Co., Ca. 1914

29 *Stanley Tools, Catalogue No. 34,* Stanley Rule & Level Co., 1914

30 *Rules, Planes, Gauges, Plumbs and Levels, Hand Screws, Handles, Spoke Shaves, Box Scrapers, Etc., Catalogue No. 114,* The Chapin-Stephens Co., 1914 (reprinted by Ken Roberts Publishing Co.)

31 *Stanley Tools, Catalogue No. 34,* Stanley Rule & Level Co., 1915

32 *Stanley Tools, Catalogue No. 34,* Stanley Rule & Level Co., 1917

33 *Carpenters and Mechanics Tools, Catalogue No. 120,* Stanley Rule & Level Co., 1919

34 *Carpenters and Mechanics Tools, Catalogue No. 34,* The Stanley Rule & Level Plant of The Stanley Works, 1922

35 *Tools for Coopers and Gaugers (and) Produce Triers, Catalogue No. 20,* A.F. Brombacher & Co., Inc., 1922 (reprinted)

36 *Carpenters and Mechanics Tools, Catalogue No. 120,* Stanley Rule & Level Co., 1923

37 *Mechanics Tool Catalogue, Catalogue No. 43,* George A. Rubelman Hardware Co., 1923

38 *Stanley Tools for Carpenters and Mechanics, Catalogue No. 34,* The Stanley Rule & Level Plant of The Stanley Works, 1925

39 *Stanley Tools for Carpenters and Mechanics, Catalogue No. 34,* The Stanley Rule & Level Plant of The Stanley Works, 1926

40 *Stanley Tools for Carpenters and Mechanics, Catalogue No. 34,* The Stanley Rule & Level Plant of The Stanley Works, 1927 plus 1929 Supplement

41 *Stanley Tools for Carpenters and Mechanics, Catalogue No. 129,* The Stanley Rule & Level Plant of The Stanley Works, 1929 (Reprinted by Roger K. Smith)

42 *Stanley Tools for Carpenters and Mechanics, Catalogue No. 34,* The Stanley Rule & Level Plant of The Stanley Works, 1929 plus 1930 Supplement

43 *Stanley Tools for Carpenters and Mechanics, Catalogue No. 34,* The Stanley Rule & Level Plant of The Stanley Works, 1929 plus 1932 Supplement

44 *Stanley Tools for Carpenters and Mechanics, Catalogue No. 134,* The Stanley Rule & Level Plant of The Stanley Works, 1934

45 *Stanley Tools for Carpenters and Mechanics, Catalogue No. 34,* The Stanley Rule & Level Plant of The Stanley Works, July 1936

46 *Stanley Tools for Carpenters and Mechanics, Catalogue No. 138,* The Stanley Rule & Level Plant of The Stanley Works, 1937

47 *Stanley Tools for Carpenters and Mechanics, Catalogue No. 138,* The Stanley Rule & Level Plant of The Stanley Works, 1938

48 *Stanley Tools for Carpenters and Mechanics, Catalogue No. 34,* The Stanley Rule & Level Plant of The Stanley Works, 1936 plus April 1938 Supplement

49 *Stanley Tools, Catalogue No. 34,* Stanley Tools Division of The Stanley Works, May 1940

50 *Stanley Tools, Catalogue No. 34,* Stanley Tools Division of The Stanley Works, January 1941

51 *Stanley Tools, Catalogue No. 34,* Stanley Tools Division of The Stanley Works, August 1941

52 *Price List of Stanley Tools* (Supplement to July 1941 Dealers Catalogue), Stanley Tools Division of The Stanley Works, December 1942

53 *Price List of Stanley Tools* (Supplement to July 1941 Dealers Catalogue), Stanley Tools Division of The Stanley Works, November 1943

54 *Stanley Tools, Catalogue No. 34,* Stanley Tools Division of The Stanley Works, 1947

55 *Stanley Tools, Catalogue No. 34,* Stanley Tools Division of The Stanley Works, 1949

56 *Lufkin Tapes, Rules, & Precision Tools, Catalogue No. 12-C,* The Lufkin Rule Co., Date Unknown

57 *Combined Price Book,* Stanley Tools Division of The Stanley Works, October 6, 1950

58 *Stanley Tools, Catalogue No. 34,* Stanley Tools Division of The Stanley Works, 1955

59 *Combined Price Book,* Stanley Tools Division of The Stanley Works, March 11, 1957

60 *Stanley Tools, Catalogue No. 34,* Stanley Tools Division of The Stanley Works, 1958

61 *Combined Price Book,* Stanley Tools Division of The Stanley Works, January 22, 1960

62 *Stanley Tools, List Price Book,* Stanley Tools Division of The Stanley Works, January 1961

63 *Stanley Tools, Dealer Price List No. PL163,* Stanley Tools Division of The Stanley Works, January 1963

64 *Stanley Tools, Catalogue No. 34,* Stanley Tools Division of The Stanley Works, 1964

65 *Stanley Tools, Dealer Price List No. PL866,* Stanley Tools Division of The Stanley Works, August 1966

66 *Stanley Tools, Dealer Price List No. PL568,* Stanley Tools Division of The Stanley Works, May 1968

67 *Stanley Tools, Broad Line Catalogue No. 34,* Stanley Tools Division of The Stanley Works, 1971/1972

68 *Stanley Tools, Broadline Catalogue No. 34,* Stanley Tools Division of The Stanley Works, 1975

Other Material

101 Coggeshall, Henry, *Timber Measure by a Line of More Ease, Dispatch, and Exactness Than Any Other Way Now in Use, By a Double Scale ...* , London, 1677

102 Bion, Nicholas, *The Construction and Principal Uses of Mathematical Instruments,* Paris, 1723, Translated by Edmond Stone, ... *To Which Are Added, the Construction and Uses of Such Instruments As Are Omitted by M. Bion,* by Edmond Stone; The Second Edition, ... *To Which is Added, a Supplement Containing a Further Account of Some of the Most Useful Mathematical Instruments As Now Improved,* Supplement by Edmond Stone, J. Richardson, Publisher, London, 1758

103 *Instructions for the Engineer's Sliding Rule, With Examples of Its Application,* Hermon Chapin, Publisher, 1858

104 "New Britain Manufactures, No. 7 – The Stanley Rule and Level Co.", *The New Britain Record,* May 18, 1866

105 Scribner, J.M., *Scribner's Ready Reckoner and Log Book* (4th Edition), George W. Fisher, Publisher, 1866

106 Rabone, John Jr., "Measuring Rules", *Birmingham and the Midland Hardware District, the Resources, Products, and Industrial History,* pp. 628-632, Samuel Timmins, Editor, London, 1866 (included as an addendum to the reprinted 1892 Catalogue of John Rabone & Sons [Ref. 16])

107 Routledge, Josiah, *Instructions for the Use of the Practical Engineers' & Mechanics' Improved Slide Rule, as Arranged By J. Routledge, Engineer,* John Rabone and Son, Publisher, Birmingham, Ca. 1867 (Reprinted by Ken Roberts Publishing Co.)

108 *Atlas of Hartford County, Connecticut,* Baker & Tilden, Publishers, Hartford, 1869

109 "Rules", *The Great Industries of the United States,* pp. 739-743, J.B. Burr & Hyde, Publishers, Hartford, 1873

110 *The Encyclopedia Britannica* (9th Edition), Charles Scribner's Sons, Publishers, New York, 1878

111 *The Carpenter's Slide Rule, Its Hisotry and Use* (3rd Edition), John Rabone & Sons, Publishers, Birmingham, 1880 (Reprinted by Ken Roberts Publishing Co.)

112 *Wood-Working Tools; How to Use Them,* Ginn, Heath & Co., Boston, 1884

113 Hogg, James, *A Practical Course of Instruction With Hogg's Improved Slide Rule in Arithmetic and Mensuration; ... (and) ... in the Calculations of Machinery for the Superintendent and Overseer in the Textile Manufacture, With Rules, Calculations, and Tables,* Rand, Avery Company, Publisher, Boston, 1887

114 Chase, I. McKim, ME, *The Art of Pattern-Making,* John Wiley & Sons, New York, 1903

115 Cajori, Florian, *A History of the Logarithmic Slide Rule, The Engineering News Publishing Co.,* New York, 1909

116 Drake, H.E., *Rule Manufacturing,* (private papers), 1910 (reprinted in *Wood Planes in 19th Century America, Vol. II* [Ref. 145])

117 Crussell, E.H., *Jobbing Work for the Carpenter,* David Williams Co., New York, 1914

118 *Machinery's Encyclopedia,* Erik Oberg & Franklin D. Jones, Editors, The Industrial Press, New York, 1917

119 Walter H.S., "History of the Rule Shop", *The Rule Shop Seismograph,* I:5 (June 16, 1919), Stanley Rule & Level Co., New Britain

120 "The Story of the Steel Square", *The Rule Shop Seismograph,* I:9 (October 15, 1919), Stanley Rule & Level Co., New Britain

121 "Trade Mark for Stanley Works and Stanley Rule & Level Plants Adopted", *The Rule Shop Seismograph,* II:9 (October 15, 1920), Stanley Rule & Level Plant of The Stanley Works, New Britain

122 Graham, Frank D., and Emery, Thomas J., *Audel Carpenters and Builders Guide,* Theodore Audel & Co., New York, 1923

123 Sanford, Vera, *A Short History of Mathematics,* Houghton Mifflin Co., Boston, 1930

124 "I Supervise the Making of Boxwood Rules", *Tool Talks, No. 11,* Stanley Tools Division of The Stanley Works, 1938

125 Townsend, Gilbert, S.B., *Steel Square,* American Technical Society, Publisher, New York, 1939

126 Chapelle, Howard I., *Boatbuilding,* W.W. Norton Co., New York, 1941

127 Benedict, Otis, Jr., *Manual of Foundry and Pattern Shop Practise,* McGraw-Hill Book Co., New York, 1947

128 Leavitt, Robert K., "History of The Stanley Works", *Stanley World,* (1951-1956), The Stanley Works, New Britain

129 Samans, Carl H., *Engineering Metals and Their Alloys,* The MacMillan Co., New York, 1953

130 Goodman, W.L., *The History of Woodworking Tools,* David McKay Company, New York, 1964

131 Freese, Frank, *A Collection of Log Rules,* General Technical Report No. FPL 1, US Department of Agriculture Forest Service, 1974

132 Watson, Aldren A., *The Village Blacksmith,* Thomas Y. Crowell Co., New York, 1977

133 Keller, David, "Lufkin, 1869 to 1977", *Cooper Industries, 1833-1977, An American History,* Cooper Industries, Inc., Publisher, Houston, 1977

134 Townsend, Raymond R., "The Ivory Rule," *The Spinning Wheel,* (September 1978), The American Antiques & Crafts Society

135 Horsley, John E., *Tools of the Maritime Trades,* International Marine Publishing Co., Camden, 1979

136 Salaman, R.A., *Dictionary of Tools Used in the Woodworking and Allied Trades,* Charles Scribner's Sons, New York, 1979

137 Kebabian, Paul B., "The Engineer's Rule", *The Active Scrapbook,* No. 30 (July 1979)

138 Kebabian, Paul B., "Stearns Rules", *The Active Scrapbook,* No. 35 (January 1981)

139 Smith, Roger K., *Patented Transitional & Metallic Planes In America,* 1827-1927, The North Village Publishing Co., Lancaster, 1981

140 Roberts, Kenneth D., P.E., "The L.C. Stephens Patent Combination Rule, Square, Level, and Bevel", *Chronicle of the Early American Industries Association,* 35:2 (June 1982), p. 29

141 West, Robert (Manager of Product Testing & Research [retired], Stanley Tools Division of The Stanley Works), *Private Communication,* 1982

142 Roberts, Kenneth D., P.E., *Introduction to Rule Collecting,* Ken Roberts Publishing Co., Fitzwilliam, 1982

143 Roberts, Kenneth D., P.E., "Carpenter's and Engineer's Slide Rules (Part I, History)", *Chronicle of the Early American Industries Association,* 36:1 (March 1983), pp. 1-5

144 *Lufkin Measuring Instruments; Excerpts From Trade Catalogues, 1888 to 1940,* Kenneth D. Roberts, Editor, Ken Roberts Publishing Co., Fitzwilliam, 1983

145 Roberts, Kenneth D., P.E., *Wood Planes in 19th Century America, Vol. II, Planemaking by the Chapins At Union Factory, 1826-1929,* Ken Roberts Publishing Co., Fitzwilliam, 1983

146 Stanley, Philip E., "Carpenters' and Engineers' Slide Rules (Part II, Routledge's Rule)", *Chronicle of the Early American Industries Association,* 1984 (to be published)

147 Stanley, Philip E., "Carpenters' and Engineers' Slide Rules (Part III, Errors in the Data Tables)", *Chronicle of the Early American Industries Association,* 1984 (to be published)

148 Hill, Alexander James, III, *Measures & Makers,* Alexander J. Hill Publishing Co., Gaithersburg, 1984 (to be published)